AVITOPIA

Birds of Jamaica

Wolfgang J. Daunicht

Cover image Streamertail, Photo: W.J.Daunicht [AU]
Copyright © AVITOPIA, Wolfgang J. Daunicht, 2021

All rights in this E-book are held by AVITOPIA, Wolfgang J. Daunicht. Licence terms applying to the individual images are placed at the end of the E-book.

AVITOPIA, Prof. Dr. Wolfgang J. Daunicht
Max-Born-Straße 12, D-60438 Frankfurt am Main
Telefax: +49(69)90756638
E-Mail: admin@avitopia.de
m36j7aoudr802k23qrllgknr14

Further information is available at

www.avitopia.net

Table of Contents

Preface	5
Bird Topography	7
Species of Birds	8
Ducks and Geese	8
Guineafowl	12
Grebes	13
Flamingoes	14
Petrels	14
Storm-petrels	16
Tropic-birds	17
Storks	18
Frigate-birds	18
Gannets, Boobies	19
Cormorants	20
Darters (Anhingas)	21
Pelicans	21
Herons	22
Ibises	25
New World Vultures	26
Ospreys	27
Birds of Prey	27
Rails, Waterhens, and Coots	29
Limpkins	31
Stilts and Avocets	32
Oystercatcher	33
Plovers	33
Jacanas	35
Sandpipers and Snipes	36
Jaegers	42
Gulls	43
Pigeons and Doves	48
Cuckoos	52
Barn owls	54
Owls	54
Nightjars	56
Potoos	57
Swifts	58
Hummingbirds	59
Todis	60
Kingfishers	61
Woodpeckers	61
Falcons	62
New World and African Parrots	63
Tyrant-flycatchers	65
Tityras and allies	68
Vireos and Shrike-Babblers	69

Ravens	70
Swallows	71
Gnatcatchers	73
Thrushes	74
Mockingbirds	76
Starlings	77
Wagtails and Pipits	78
Waxwings	78
American Warblers	79
Tanagers	87
New World Buntings and Sparrows	89
Spindalises	89
Cardinals	90
New World Blackbirds	92
Finches	94
Sparrows	94
Weavers	95
Waxbills	95
Index of English Names	97
Index of Scientific Names	100
Additional Copyright Terms	104

Preface

Jamaica is a souvereign State.

Jamaica has an amazing variety of landscapes - from the wetlands along the coasts to the slow rivers, to the torrents to the mountainous rainforests. This diversity is reflected in a fascinating birdlife with many endemic species. Therefore BirdLife International has designated Jamaica as an EBA-P (Primary Endemic Bird Area) and also registered 16 IBAs (Important Bird and Biodiversity Areas), i.e. sites of global importance for the conservation of bird life.

A list of the IBAs (Important Bird and Biodiversity Areas) of this country is available in the Data Zone of http://datazone.birdlife.org/site/results?cty=106. In this list you can call up a map and further information for each IBA with one click.

The assessment of the global conservation status of bird species uses the criteria of the Red List (IUCN) 2012.

Legend
△ Near threatened
▲ Vulnerable
▲ Endangered
▲ Critically endangered
▲ Extinct in the wild
▲ Data deficient
Ø Invalid taxon
† Extinct
(e) Picture of an endemic subspecies
🔊 Link to video with audio
▢ Link to video
🔊 Link to audio

This e-book is based on a request to the AVITOPIA Data Base the 04th July 2021.
The request profile was:
- Primary language: English - Secondary language: unrestricted
- Maximum number of pictures per species: 1
- Content and illustration: all names, optimal illustration
- Scientific system: Clements et al. 2017
- Method of area selection: Menu tree
- Name of area: Jamaica
- Survival criterion: unrestricted
- Selection of a taxon: all birds
- Taxonomic depth: Species of Birds
- Selection of activity/nest/portrait: unrestricted
- Selection of plumage/egg(s): unrestricted
- Selection of image technique: unrestricted

In the resulting PDF or ePub file, resp., all index and register entries are linked.

Bird Topography

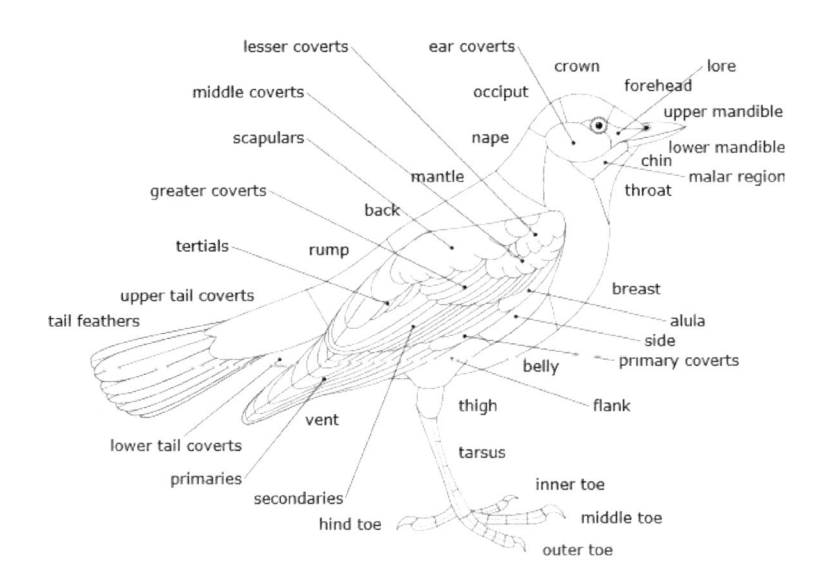

Species of Birds

Ducks and Geese - *Anatidae*

The family of Ducks and Geese occurs in all continents of the world except in Antarctica. The birds grow up to 30 - 180 cm long and live essentially on the water. The front three toes are webbed, the fourth toe is small and shifted upwards. All species swim, some dive well. Most species fly well, only a few are flightless. However, shortly after the breeding season, the birds adopt simple plumage and shed all flight feathers, so that they are unable to fly for some time. The nests are very diverse: there are nests on the ground, in ground caves, in steep walls and in tree hollows. The clutch comprises 4 to 12 eggs, the incubation lasts between 3 and 5 weeks and the young leave the nest soon after hatching.

West Indian Whistling Duck
  de: Kubapfeifgans
  fr: Dendrocygne des Antilles
  es: Suiriń Yaguaza
  ja: ハシグロリュウキュウガモ
  cn: 西印度树鸭
Dendrocygna arborea
Vulnerable.

Fulvous Whistling Duck
  de: Gelbbrust-Pfeifgans
  fr: Dendrocygne fauve
  es: Suiriń Bicolor
  ja: アカリュウキュウガモ
  cn: 茶色树鸭
Dendrocygna bicolor

www.avitopia.net/bird.en/?vid=200106

Wood Duck
  de: Brautente
  fr: Canard branchu
  es: Pato Joyuyo
  ja: アメリカオシ
  cn: 林鸳鸯
Aix sponsa

Blue-winged Teal
de: Blauflügelente
fr: Sarcelle à ailes bleues
es: Cerceta Aliazul
ja: ミカヅキシマアジ
cn: 蓝翅鸭
Spatula discors

Northern Shoveler
de: Löffelente
fr: Canard souchet
es: Cuchara Común
ja: ハシビロガモ
cn: 琵嘴鸭
Spatula clypeata

www.avitopia.net/bird.en/?vid=203010

Gadwall
de: Schnatterente
fr: Canard chipeau
es: Anade Friso
ja: オカヨシガモ
cn: 赤膀鸭
Mareca strepera

www.avitopia.net/bird.en/?vid=203101

American Wigeon
de: Nordamerikanische Pfeifente
fr: Canard d'Amérique
es: Silbón Americano
ja: アメリカヒドリ
cn: 绿眉鸭
Mareca americana

White-cheeked Pintail
de: Bahamaente
fr: Canard des Bahamas
es: Anade Gargantillo
ja: ホオジロオナガガモ
cn: 白脸针尾鸭
Anas bahamensis

www.avitopia.net/bird.en/?vid=203214

Northern Pintail
de: Spießente
fr: Canard pilet
es: Ánade Rabudo
ja: オナガガモ
cn: 针尾鸭
Anas acuta

www.avitopia.net/bird.en/?vid=203216

Green-winged Teal
de: Krickente
fr: Sarcelle d'hiver
es: Cerceta común
ja: コガモ
cn: 绿翅鸭
Anas crecca

www.avitopia.net/bird.en/?vid=203219

Canvasback
de: Riesentafelente
fr: Fuligule à dos blanc
es: Porrón Coacoxtle
ja: オオホシハジロ
cn: 帆背潜鸭
Aythya valisineria

Redhead
  de: Rotkopfente
  fr: Fuligule à tête rouge
  es: Porrón Americano
  ja: アメリカホシハジロ
  cn: 美洲潜鸭
Aythya americana

♂ adult

Ring-necked Duck
  de: Ringschnabelente
  fr: Fuligule à collier
  es: Porrón Acollarado
  ja: クビワキンクロ
  cn: 环颈潜鸭
Aythya collaris

♂ adult

Lesser Scaup
  de: Kleine Bergente
  fr: Petit Fuligule
  es: Porrón Bola
  ja: コスズガモ
  cn: 小潜鸭
Aythya affinis

♂ adult

Hooded Merganser
  de: Kappensäger
  fr: Harle couronné
  es: Serreta Capuchona
  ja: オウギアイサ
  cn: 棕胁秋沙鸭
Lophodytes cucullatus

♂ adult

Masked Duck
  de: Maskenruderente
  fr: Érismature routoutou
  es: Malvasía Enmascarada
  ja: メンカブリオタテガモ
  cn: 花脸硬尾鸭
Nomonyx dominicus

♂ adult

Ruddy Duck
  de: Schwarzkopf-Ruderente
  fr: Érismature rousse
  es: Pato Malvasia de Cara Blanca
  ja: アカオタテガモ
  cn: 棕硬尾鸭
Oxyura jamaicensis

♂ adult

Guineafowl - *Numididae*

The family of Guinefowls is found in sub-Saharan Africa and Madagascar. The body length is 43 cm to 74 cm. Guinea fowl are plump birds with strong legs. The beak is short and strong, the hind toe is shifted upwards. The birds are ground dwellers, although they can fly quite well. When they flee, they run and don't fly. In contrast, they usually spend the night in trees. They eat almost all kinds of parts of plants, but also insects and other animals. The nest consists of a dug out in the ground in which up to 20 eggs are laid. The chicks are precocial and looked after by both parents.

Helmeted Guineafowl
  de: Helmperlhuhn
  fr: Pintade de Numidie
  es: Pintada Común
  ja: ホロホロチョウ
  cn: 珠鸡
Numida meleagris
Introduced.

https://www.avitopia.net/bird.en/?vid=275101
https://www.avitopia.net/bird.en/?aud=275101

adult

Grebes - *Podicipedidae*

The family of Grebes are found on freshwater lakes around the world, except in the extreme north and south and on some islands. In winter they can also be found on the coast of the sea. The size ranges from 20 cm to 50 cm, the wings are short, tail feathers are missing. The toes have flap-like widenings. They only fly regularly and at night during the migration time. In addition, they are well adapted to aquatic life. Both parents lead the striped or spotted young birds until they become independent.

Least Grebe
- de: Schwarzkopftaucher
- fr: Grèbe minime
- es: Zampullín Macacito
- ja: ヒメカイツブリ
- cn: 侏□□

Tachybaptus dominicus

breeding

Pied-billed Grebe
- de: Bindentaucher
- fr: Grèbe à bec bigarré
- es: Zampullín Picogrueso
- ja: オビハシカイツブリ
- cn: 斑嘴巨□□

Podilymbus podiceps

adult

Flamingoes - *Phoenicopteridae*

The family of Flamingos includes only a few species, but they have spread to the warm areas of all continents except Australia. With their long legs and long necks, the birds reach heights of up to 120 cm. The wings are large, the tail is short. The beaks are unique: bent down in the middle with a channel-shaped lower beak and a lid-shaped upper beak. In the water the beak is held upright; the tongue acts as a piston that sucks in the water and presses it out through the filters of the beak. Flamingos feed on small animals and parts of plants that float in the water. They breed in colonies, the nests are truncated cones made of mud. The young birds flee the nest and are fed by both parents.

American Flamingo
de: Kubaflamingo
fr: Flamant des Caraïbes
es: Flamenco
ja: オオフラミンゴ
cn: 大红鹳
Phoenicopterus ruber

adult

Petrels - *Procellariidae*

The family of Petrels is at home at sea all over the world. They essentially use the land for breeding, and some species even do so on the coast of Antarctica. Most of the species are migratory birds. The body length ranges from 30 cm to 90 cm. The birds have long, pointed wings and short tails, and their feet are webbed. The smaller species breed in caves or crevices, they defend the young birds by vomiting stinking oil.

Black-capped Petrel
de: Teufelssturmvogel
fr: Pétrel diablotin
es: Petrel Antillano
ja: ズグロシロハラミズナギドリ
cn: 黑顶圆尾鹱
Pterodroma hasitata
Endangered.

ssp. caribbaea adult

Cory's Shearwater
 de: Gelbschnabel-Sturmtaucher
 fr: Puffin cendré
 es: Pardela cenicienta
 ja: オニミズナギドリ
 cn: 猛鹱
Calonectris diomedea

adult

Sooty Shearwater
 de: Dunkler Sturmtaucher
 fr: Puffin fuligineux
 es: Pardela Sombría
 ja: ハイイロミズナギドリ
 cn: 灰鹱
Ardenna grisea
Near threatened.

adult

Manx Shearwater
 de: Schwarzschnabel-Sturmtaucher
 fr: Puffin des Anglais
 es: Pardela Pichoneta
 ja: マンクスミズナギドリ
 cn: 大西洋鹱
Puffinus puffinus

adult

Audubon's Shearwater
 de: Audubonsturmtaucher
 fr: Puffin d'Audubon
 es: Pardela Garrapatera
 ja: セグロミズナギドリ
 cn: 奥氏鹱
Puffinus lherminieri

adult

Storm-petrels - *Hydrobatidae*

The family of Storm-petrels is widespread on all oceans of the earth and occurs partly in large numbers. With a body length of 14 cm to 25 cm, they are the smallest seabirds with webbed feet. Your weak legs are hardly able to carry their body weight without the support of the wings. They breed in colonies in caves or crevices, which they usually only attend at night. Although they usually breed on mammal-free islands, the greatest danger comes from introduced mammals. The Guadalupe storm-petrel was driven to extinction by feral cats.

Wilson's Storm Petrel
de: Buntfuß-Sturmschwalbe
fr: Océanite de Wilson
es: Paíño de Wilson
ja: アシナガウミツバメ
cn: 黄蹼洋海燕
Oceanites oceanicus

Leach's Storm Petrel
de: Wellenläufer
fr: Océanite cul-blanc
es: Paíño Boreal
ja: コシジロウミツバメ
cn: 白腰叉尾海燕
Oceanodroma leucorhoa

Tropic-birds - *Phaethontidae*

The family of Tropicbirds occurs on all tropical oceans. With a length of 41 cm to 48 cm, they belong to the medium-sized birds. They have long, pointed wings and a wedge-shaped tail, the central feathers of which are greatly elongated. The legs are very short, the feet are webbed. The head is large with a strong, slightly curved bill. They fly quickly and pounce on small fish and octopus out of the air. They nest in large, heavily populated colonies on the ground or in a crevice. The young are altricial, are fed by both parents and only fledged after up to 15 weeks.

White-tailed Tropicbird
 de: Weißschwanz-Tropikvogel
 fr: Phaéton à bec jaune
 es: Rabijunco Menor
 ja: シラオネッタイチョウ
 cn: 白尾鹲
Phaethon lepturus

adult

Red-billed Tropicbird
 de: Rotschnabel-Tropikvogel
 fr: Phaéton à bec rouge
 es: Rabijunco Etéreo
 ja: アカハシネッタイチョウ
 cn: 红嘴鹲
Phaethon aethereus

adult

Storks - *Ciconiidae*

The family of storks is widespread worldwide except for the coldest areas. Some species are resident birds, others are long-distance migrants. The body length ranges from 75 cm to 150 cm. They are long-legged birds with large wings, a long neck and a long beak. They fly a lot, usually with a stretched neck, and they sail excellently. They feed on small animals, from insects to small mammals. 3 to 6 eggs are laid in the shallow nest made of brushwood, which are incubated by both parents.

Wood Stork
de: Waldstorch
fr: Tantale d'Amérique
es: Tántalo Americano
ja: アメリカトキコウ
cn: 黑头鹮鹳
Mycteria americana

adult

Frigate-birds - *Fregatidae*

The family of Frigate-birds can be found at all tropical and subtropical seas. Their length ranges from 80 cm to 105 cm. They have very long wings, a curved beak, and a forked tail. The males have a scarlet throat pouch during the breeding season, which is inflated during courtship. They feed on fish, but never go down on the water, instead catch them in flight. Frigate birds often parasitize other birds by chasing them until they regurgitate their food. The nest can reach a diameter of 4.5 m. The single egg is incubated by both adult birds for 40 days. The chick is fledged after 4 to 5 months.

Magnificent Frigatebird
de: Prachtfregattvogel
fr: Frégate superbe
es: Rabihorcado Magnífico
ja: アメリカグンカンドリ
cn: 华丽军舰鸟
Fregata magnificens

♂ adult

Gannets, Boobies - *Sulidae*

The family of Boobies is common on all seas near the coast. The birds are 65 cm to 100 cm long. The wings are long and pointed, the legs are short and the feet are large and webbed. The beak is strong and has no nostrils. Boobies are extremely elegant fliers, but quite awkward on the ground. They hunt fish for which they plunge into water from a height of up to 35 m in order to pursue them under water and to grab them with their beak. They breed in colonies on the ground or on trees. The young birds are provided with regurgitated food.

Masked Booby
 de: Maskentölpel
 fr: Fou masqué
 es: Alcatraz Enmascarado
 ja: アオツラカツオドリ
 cn: 蓝脸鲣鸟
Sula dactylatra

Brown Booby
 de: Weißbauchtölpel
 fr: Fou brun
 es: Piquero Pardo
 ja: カツオドリ
 cn: 褐鲣鸟
Sula leucogaster

Red-footed Booby
 de: Rotfußtölpel
 fr: Fou à pieds rouges
 es: Piquero Patirrojo
 ja: アカアシカツオドリ
 cn: 红脚鲣鸟
Sula sula

Northern Gannet
de: Basstölpel
fr: Fou de Bassan
es: Alcatraz Atlántico
ja: シロカツオドリ
cn: 北鲣鸟
Morus bassanus

♂♀ adult

Cormorants - *Phalacrocoracidae*

The cormorant family are gregarious freshwater or marine birds found on every continent. The body length is 50 cm to 100 cm. They have short legs with large webbed feet. They have a long neck and a slender beak with a curved tip. They dive from the surface of the water and can stay underwater for minutes. The caught fish are thrown into the air and devoured head first.

Neotropic Cormorant
de: Olivenscharbe
fr: Cormoran vigua
es: Cormorán Biguá
ja: ナンベイヒメウ
cn: 美洲鸬鹚
Phalacrocorax brasilianus

adult

Darters (Anhingas) - *Anhingidae*

The family of Darters occurs in the tropics and subtropics of all continents. The northern populations are migratory birds. The body size is 90 cm. The birds have long wings, a long trunk and tail. The legs are short, the large feet are webbed. The head is small and the neck and pointed beak are long and slender. They use their pointed beak to spear fish when hunting underwater. They cannot grease their feathers, so they soak up water and have to be dried with outstretched wings. Darters nest in colonies often in the company of other water birds. The male chooses the nesting site and carries branches that the female uses to build the nest.

Anhinga
  de: Amerikanischer Schlangenhalsvogel
  fr: Anhinga d'Amérique
  es: Anhinga Americana
  ja: アメリカヘビウ
  cn: 美洲蛇鹈
Anhinga anhinga

♂ adult

Pelicans - *Pelecanidae*

The family of Pelicans is scattered across all continents. The body length is 125 cm to 180 cm. The legs are short and strong, the toes are webbed. The wings are large, the tail is short, and the birds sail excellently. The beak is long, straight and flat, and at the lower part of the beak is a large, stretchy pouch that is used like a fishing net. Pelicans are very sociable and often work together to fish. Some species plunge into the water from great heights. Pelicans nest in colonies.

Brown Pelican
  de: Braunpelikan
  fr: Pélican brun
  es: Pelícano Pardo
  ja: カッショクペリカン
  cn: 褐鹈鹕
Pelecanus occidentalis

adult

Herons - *Ardeidae*

The heron family of Herons occurs on all continents and on many islands. The body length ranges from 28 nm to 140 cm. The wings are large, the tail is short. Legs, toes and neck are long, the latter has a characteristic S-shape. They feed mainly on fish, but also eat other small animals. They mostly breed in colonies. The food brought in is regurgitated in front of the chicks.

American Bittern
- de: Nordamerikanische Rohrdommel
- fr: Butor d'Amérique
- es: Avetoro Lentiginoso
- ja: アメリカサンカノゴイ
- cn: 美洲麻鳽

Botaurus lentiginosus

Least Bittern
- de: Amerikanische Zwergdommel
- fr: Petit Blongios
- es: Avetorillo Panamericano
- ja: コヨシゴイ
- cn: 姬苇鳽

Ixobrychus exilis

Great Blue Heron
- de: Kanadareiher
- fr: Grand Héron
- es: Garza Azulada
- ja: オオアオサギ
- cn: 大蓝鹭

Ardea herodias

Great Egret
- de: Silberreiher
- fr: Grande Aigrette
- es: Garceta Grande
- ja: ダイサギ
- cn: 大白鷺

Ardea alba

www.avitopia.net/bird.en/?vid=750711

adult

Snowy Egret
- de: Schmuckreiher
- fr: Aigrette neigeuse
- es: Garceta Nívea
- ja: ユキコサギ
- cn: 雪鷺

Egretta thula

adult

Little Blue Heron
- de: Blaureiher
- fr: Aigrette bleue
- es: Garceta Azul
- ja: ヒメアカクロサギ
- cn: 小蓝鹭

Egretta caerulea

adult

Tricolored Heron
- de: Dreifarbenreiher
- fr: Aigrette tricolore
- es: Garceta Tricolor
- ja: サンショクサギ
- cn: 三色鷺

Egretta tricolor

adult

Reddish Egret
de: Rötelreiher
fr: Aigrette roussâtre
es: Garceta Rojiza
ja: アカクロサギ
cn: 棕頸鷺
Egretta rufescens
Near threatened.

adult, dark phase

Cattle Egret
de: Kuhreiher
fr: Héron garde-boeufs
es: Garcilla Bueyera
ja: アマサギ
cn: 牛背鷺
Bubulcus ibis

www.avitopia.net/bird.en/?vid=750901

breeding

Green Heron
de: Grünreiher
fr: Héron vert
es: Garcita verdosau
ja: アメリカササゴイ
cn: 美洲绿鹭
Butorides virescens

adult

Black-crowned Night Heron
de: Nachtreiher
fr: Bihoreau gris
es: Martinete Común
ja: ゴイサギ
cn: 夜鷺
Nycticorax nycticorax

www.avitopia.net/bird.en/?vid=751501

breeding

Yellow-crowned Night Heron
- de: Krabbenreiher
- fr: Bihoreau violacé
- es: Martinete Coronado
- ja: シラガゴイ
- cn: 黄冠夜鹭

Nyctanassa violacea

adult

Ibises - *Threskiornithidae*

The family of Ibises occurs in all warm regions of the world. The birds are 50 cm to 110 cm high. They have long wings and a short tail. The toes are connected by small webs. The long beak is either curved downwards or broad and spatulate. Most species are quite gregarious. They fly powerfully and can sail with the neck stretched out. Their food is very varied. The Sacred Ibis was the revered symbol of the god Thoth in ancient Egypt, but is now extinct in this country.

American White Ibis
- de: Schneesichler
- fr: Ibis blanc
- es: Corocoro Blanco
- ja: シロトキ
- cn: 美洲白鹮

Eudocimus albus

adult

Glossy Ibis
- de: Sichler
- fr: Ibis falcinelle
- es: Morito Común
- ja: ブロンズトキ
- cn: 彩鹮

Plegadis falcinellus

www.avitopia.net/bird.en/?vid=775201

adult

Roseate Spoonbill
de: Rosalöffler
fr: Spatule rosée
es: Espátula Rosada
ja: ベニヘラサギ
cn: 粉红琵鹭
Platalea ajaja

breeding

New World Vultures - *Cathartidae*

The family of New World Vultures is widespread in America, northward to southern Canada. The body length ranges from 65 cm to 110 cm. With up to 3.20 m, the wingspan is among the largest among birds. The birds have strong hooked bills, a featherless head and neck, and weak claws. They soar and glide well often at great heights. Their sense of smell is good and the New World Vultures use it to find food. They live mainly on carrion, rarely on live prey. One or two eggs are laid, which are incubated by both partners.

Turkey Vulture
de: Truthahngeier
fr: Urubu à tête rouge
es: Aura Gallipavo
ja: ヒメコンドル
cn: 红头美洲鹫
Cathartes aura

adult

Ospreys - *Pandionidae*

There is only one species in the family of ospreys, but it is a true cosmopolitan: it occurs on all continents. The birds are around 60 cm tall with long wings and short tails. The beak is hook-shaped and the feet have warty sole pads. Fischdler are pure fish hunters who hover first and then plunges into the water, often submerging themselves completely. The fish are grasped with the claws and carried to a sitting site or a nest. The female breeds and cares for the young birds alone.

Osprey
  de: Fischadler
  fr: Balbuzard pêcheur
  es: Águila Pescadora
  ja: ミサゴ
  cn: 鹗
Pandion haliaetus

♀ adult

Birds of Prey - *Accipitridae*

The family of Birds of Prey is found worldwide with the exception of the Arctic, Antarctic and most of the oceanic islands. Birds of Prey are of various sizes (20 - 115 cm), they have long, round wings, curved claws and a short hooked bill. The sexes are almost the same, the female is almost always larger. All species are good fliers, and many sail well too. They mainly hunt live animals, only the vultures are scavengers. Even fishing species are among them.

Swallow-tailed Kite
  de: Schwalbenweih
  fr: Milan à queue fourchue
  es: Elanio Tijereta
  ja: ツバメトビ
  cn: 燕尾鸢
Elanoides forficatus

adult

Mississippi Kite
de: Mississippiweih
fr: Milan du Mississippi
es: Elanio del Mississipí
ja: ミシシッピアトビ
cn: 密西西比灰鸢
Ictinia mississippiensis

adult

American Hen-harrier
de: Hudsonweihe
fr: Busard d'Amérique
es: Aguilucho de Hudson
ja: アメリカチュウヒ
cn: 北鹞
Circus hudsonius

♂ adult

Sharp-shinned Hawk
de: Eckschwanzsperber
fr: Épervier brun
es: Azor Rojizo
ja: アシボソハイタカ
cn: 纹腹鹰
Accipiter striatus

adult

Broad-winged Hawk
de: Breitflügelbussard
fr: Petite Buse
es: Busardo Aliancho
ja: ハネビロノスリ
cn: 巨翅鵟
Buteo platypterus

adult

Red-tailed Hawk
de: Rotschwanzbussard
fr: Buse à queue rousse
es: Busardo Colirrojo
ja: アカオノスリ
cn: 红尾鵟
Buteo jamaicensis

adult

Rails, Waterhens, and Coots - *Rallidae*

The family of Rails and Coots occurs worldwide except in the polar regions. Rails are at most medium-sized birds (15 - 50 cm) with short wings and very short tails. The toes are long and have swimming lobes in the coots. The sexes usually look the same. Almost all species swim well. Many only become active at dusk or are night birds. Some are able to fly long distances, while island species are partially flightless.

Yellow-breasted Crake
de: Gelbbrust-Sumpfhuhn
fr: Marouette à sourcils blancs
es: Polluela Pálida
ja: キムネヒメクイナ
cn: 黄胸田鸡
Hapalocrex flaviventer

adult

Black Rail
de: Schieferralle
fr: Râle noir
es: Burrito Negruzco
ja: クロコビトクイナ
cn: 黑田鸡
Laterallus jamaicensis
Near threatened.

adult

Clapper Rail
de: Klapperralle
fr: Râle tapageur
es: Rascón crepitante
ja: オニクイナ
cn:
Rallus crepitans

Uniform Crake
de: Einfarbralle
fr: Râle concolore
es: Cotara Café
ja: チャバラクイナ
cn: 纯色秧鸡
Amaurolimnas concolor

Sora
de: Carolinasumpfhuhn
fr: Marouette de Caroline
es: Polluela Sora
ja: カオグロクイナ
cn: 黑脸田鸡
Porzana carolina

American gallinule
de: Zwergsultanshuhn
fr: Talève violacée
es: Calamoncillo Americano
ja: アメリカムラサキバン
cn: 紫青水鸡
Porphyrio martinica

Common Gallinule
de: Amerikanisches Teichhuhn
fr: Gallinule d'Amérique
es: Gallineta americana
ja: アメリカバン
cn: 黑水鸡
Gallinula galeata

American Coot
de: Amerikanisches Blässhuhn
fr: Foulque d'Amérique
es: Focha Americana
ja: アメリカオオバン
cn: 美洲骨顶
Fulica americana

Limpkins - *Aramidae*

The family of Limpkins, which includes only one species, occurs from Florida over the Caribbean to Argentina. They grow about 70 cm long. They have broad round wings and a broad tail. Legs and toes are long with sharp claws. The beak is compressed on the side and bent slightly downwards. They live in swamps, because their main food is water snails of the genus Pomacea. They are solitary and mostly active at dusk or at night. The shallow nest is built from brushwood near the water, too. The eggs are incubated by both parents, and both adult birds also take care of the nidifugous chicks.

Limpkin
de: Rallenkranich
fr: Courlan brun
es: Carrao
ja: ツルモドキ
cn: 秧鹤
Aramus guarauna

Stilts and Avocets - *Recurvirostridae*

The family of Stilts and Avocets is widespread worldwide; the northern populations are migratory birds. The body length is 30 cm to 50 cm. They have very long legs and a thin beak that is straight or curved upwards. They fly well and can swim well. They search the mud in shallow waters for invertebrates. They nest in colonies, the nest-fleeing young birds are looked after by both parents. The defense of the offspring includes various distraction techniques from simulating a 'broken wing' to 'false brooding' in full view of a predator.

Black-necked Stilt
de: Schwarznacken-Stelzenläufer
fr: Échasse d'Amérique
es: Cigüeñuela de Cuello Negro
ja: クロエリセイタカシギ
cn: 黑颈长脚鹬
Himantopus mexicanus

www.avitopia.net/bird.en/?vid=1250104

adult

American Avocet
de: Braunhals-Säbelschnäbler
fr: Avocette d'Amérique
es: Avoceta Americana
ja: アメリカソリハシセイタカシギ
cn: 褐胸反嘴鹬
Recurvirostra americana

breeding

Oystercatcher - *Haematopodidae*

The family of oystercatchers is found in temperate and tropical waters from Iceland and the Aleutian Islands to Cape Horn and Tasmania. The body length of the medium-sized birds is 32 cm to 45 cm. The legs are long and strong, the feet have small webs. The beak is long, strong and compressed at the sides. Their diet consists of mussels, crabs, worms and insects, but oysters are not the main ingredient. Outside of the breeding season, they are sociable and then gather in large flocks that can reach a few thousand. The chicks who flee the nest are looked after by both parents until they have fledged after five weeks.

American Oystercatcher
- de: Braunmantel-Austernfischer
- fr: Huîtrier d'Amérique
- es: Ostrero Pío Americano
- ja: アメリカミヤコドリ
- cn: 美洲蛎鹬

Haematopus palliatus

adult

Plovers - *Charadriidae*

The plover family is global; many species are migratory birds. The body length ranges from 15 cm to 40 cm. Plover have a stocky body and long wings, the hind toe is receded or missing completely. They are ground birds that can run quickly, but also fly very well and quickly. In the vicinity of the nest or the young birds, the adult birds use seduction by simulating a broken wing and luring away a dangerous animal.

Grey Plover
- de: Kiebitzregenpfeifer
- fr: Pluvier argenté
- es: Chorlito Gris
- ja: ダイゼン
- cn: 灰鸻

Pluvialis squatarola

breeding

American Golden-Plover
de: Amerikanischer Goldregenpfeifer
fr: Pluvier bronzé
es: Chorlito Dorado Americano
ja: ムナグロ
cn: 美洲金鸻
Pluvialis dominica

breeding

Snowy plover
de: Schneeregenpfeifer
fr: Pluvier neigeux
es: Chorlitejo nivoso
ja: シロチドリ
cn: 雪鸻
Charadrius nivosus
Near threatened.

adult

Wilson's Plover
de: Wilsonregenpfeifer
fr: Pluvier de Wilson
es: Chorlitejo Piquigrueso
ja: ウイルソンチドリ
cn: 厚嘴鸻
Charadrius wilsonia

adult

Semipalmated Plover
de: Amerikanischer Sandregenpfeifer
fr: Pluvier semipalmé
es: Chorlitejo Semipalmeado
ja: ミカズキチドリ
cn: 半蹼鸻
Charadrius semipalmatus

non-breeding

Piping Plover
- de: Flötenregenpfeifer
- fr: Pluvier siffleur
- es: Chorlitejo Silbador
- ja: フエコチドリ
- cn: 笛鸻

Charadrius melodus
Near threatened.

adult

Killdeer
- de: Keilschwanz-Regenpfeifer
- fr: Pluvier kildir
- es: Chorlitejo Culirrojo
- ja: フタオビチドリ
- cn: 双领鸻

Charadrius vociferus

adult

Jacanas - *Jacanidae*

The family of Jacanas occurs from Africa through South Asia to Australia and in Central and South America. They become 15 cm to 50 cm long. Legs, toes and claws are long, the tail is usually short. Noteworthy is a thorn on the leading edge of the wing. Their habitat are the banks of lakes and swamps. They are very adept at walking over floating vegetation such as Water lily leaves. They eat small aquatic animals, but also seeds from aquatic plants. In most species, the males take care of the offspring.

Northern Jacana
- de: Gelbstirn-Blatthühnchen
- fr: Jacana du Mexique
- es: Jacana Centroamericana
- ja: アメリカレンカク
- cn: 美洲水雉

Jacana spinosa

adult

Sandpipers and Snipes - *Scolopacidae*

The family of Snipes is distributed worldwide, most of the species are migratory birds that sometimes cover great distances. The body length ranges from 13 cm to 60 cm. They usually live near water and outside the breeding season often form large flocks on the seashore. The diet consists of small invertebrates. The young birds leave the nest immediately after hatching.

Upland Sandpiper
de: Prärieläufer
fr: Maubèche des champs
es: Correlimos Batitú
ja: マキバシギ
cn: 高原鷸
Bartramia longicauda

adult

Whimbrel
de: Regenbrachvogel
fr: Courlis corlieu
es: Zarapito Trinador
ja: チュウシャクシギ
cn: 中杓鷸
Numenius phaeopus

www.avitopia.net/bird.en/?vid=1450202

adult

Hudsonian Godwit
de: Hudsonschnepfe
fr: Barge hudsonienne
es: Aguja Café
ja: アメリカオグロシギ
cn: 棕塍鷸
Limosa haemastica

breeding

Marbled Godwit
de: Marmorschnepfe
fr: Barge marbrée
es: Aguja Canela
ja: アメリカオオソリハシシギ
cn: 云斑塍鹬
Limosa fedoa

adult

Ruddy Turnstone
de: Steinwälzer
fr: Tournepierre à collier
es: Vuelvepiedras Común
ja: キョウジョシギ
cn: 翻石鹬
Arenaria interpres

breeding

Red Knot
de: Knutt
fr: Bécasseau maubèche
es: Correlimos Gordo
ja: コオバシギ
cn: 红腹滨鹬
Calidris canutus
Near threatened.

breeding

Ruff
de: Kampfläufer
fr: Combattant varié
es: Combatiente
ja: エリマキシギ
cn: 流苏鹬
Calidris pugnax

www.avitopia.net/bird.en/?vid=1450604

♂ breeding

Stilt Sandpiper
 de: Bindenstrandläufer
 fr: Bécasseau à échasses
 es: Correlimos Zancolín
 ja: アシナガシギ
 cn: 高跷鹬
Calidris himantopus

breeding

Sanderling
 de: Sanderling
 fr: Bécasseau sanderling
 es: Correlimos Tridáctilo
 ja: ミユビシギ
 cn: 三趾滨鹬
Calidris alba

non-breeding

Least Sandpiper
 de: Wiesenstrandläufer
 fr: Bécasseau minuscule
 es: Correlimos Menudillo
 ja: アメリカヒバリシギ
 cn: 美洲小滨鹬
Calidris minutilla

breeding

White-rumped Sandpiper
 de: Weißbürzel-Strandläufer
 fr: Bécasseau à croupion blanc
 es: Correlimos Culiblanco
 ja: コシジロウズラシギ
 cn: 白腰滨鹬
Calidris fuscicollis

breeding

Buff-breasted Sandpiper
 de: Grasläufer
 fr: Bécasseau roussâtre
 es: Correlimos Canelo
 ja: コモンシギ
 cn: 黄胸鹬
Calidris subruficollis
Near threatened.

adult

Pectoral Sandpiper
 de: Graubrust-Strandläufer
 fr: Bécasseau à poitrine cendrée
 es: Correlimos Pectoral
 ja: アメリカウズラシギ
 cn: 斑胸滨鹬
Calidris melanotos

breeding

Semipalmated Sandpiper
 de: Sandstrandläufer
 fr: Bécasseau semipalmé
 es: Correlimos Semipalmeado
 ja: ヒレアシトウネン
 cn: 半蹼滨鹬
Calidris pusilla
Near threatened.

breeding

Western Sandpiper
 de: Bergstrandläufer
 fr: Bécasseau d'Alaska
 es: Correlimos de Alaska
 ja: ヒメハマシギ
 cn: 西滨鹬
Calidris mauri

adult

Short-billed Dowitcher
de: Kleiner Schlammläufer
fr: Bécassin roux
es: Agujeta Gris
ja: アメリカオオハシシギ
cn: 短嘴半蹼鹬
Limnodromus griseus

breeding

Wilson's Snipe
de: Amerikanische Bekassine
fr: Bécassine de Wilson
es: Becasina
ja: アメリカシギ
cn: 美洲沙锥
Gallinago delicata

adult

Red-necked Phalarope
de: Odinshühnchen
fr: Phalarope à bec étroit
es: Falaropo Picofino
ja: アカエリヒレアシシギ
cn: 红颈瓣蹼鹬
Phalaropus lobatus

♀ breeding

Spotted Sandpiper
de: Drosseluferläufer
fr: Chevalier grivelé
es: Andarríos Maculado
ja: アメリカイソシギ
cn: 斑腹矶鹬
Actitis macularius

breeding

Solitary Sandpiper
   de: Einsamer Wasserläufer
   fr: Chevalier solitaire
   es: Andarríos Solitario
   ja: コシグロクサシギ
   cn: 褐腰草鷸
Tringa solitaria

adult

Greater Yellowlegs
   de: Großer Gelbschenkel
   fr: Grand Chevalier
   es: Archibebe Patigualdo Grande
   ja: オオキアシシギ
   cn: 大黄脚鷸
Tringa melanoleuca

adult

Willet
   de: Schlammtreter
   fr: Chevalier semipalmé
   es: Playero Aliblanco
   ja: ハジロオオシギ
   cn: 斑翅鷸
Tringa semipalmata

non-breeding

Lesser Yellowlegs
   de: Kleiner Gelbschenkel
   fr: Petit Chevalier
   es: Archibebe Patigualdo Chico
   ja: コキアシシギ
   cn: 小黄脚鷸
Tringa flavipes

adult

Jaegers - *Stercorariidae*

The family of Jaegers is native to the arctic and subarctic areas of the northern and southern hemispheres. They migrate very far and can spend indefinite time at sea. The body length ranges from 40 cm to 60 cm. Their feet are webbed and have strong claws. The beak is strong and has a curved tip. Skuas are very fast and agile fliers. They breed near bird colonies and are aggressive predators and parasites there. They chase other birds until they vomit their food.

South Polar Skua
de: Antarktikskua
fr: Labbe de McCormick
es: Págalo Polar
ja: ナンキョクオオトウゾクカモメ
cn: 灰贼鸥
Stercorarius maccormicki

adult

Pomarine Jaeger
de: Spatelraubmöwe
fr: Labbe pomarin
es: Págalo Pomarino
ja: トウゾクカモメ
cn: 中贼鸥
Stercorarius pomarinus

adult

Parasitic Jaeger
de: Schmarotzerraubmöwe
fr: Labbe parasite
es: Págalo Parásito
ja: クロトウゾクカモメ
cn: 短尾贼鸥
Stercorarius parasiticus

adult, dark phase

Long-tailed Jaeger
de: Falkenraubmöwe
fr: Labbe à longue queue
es: Págalo Rabero
ja: シロハラトウゾクカモメ
cn: 长尾贼鸥
Stercorarius longicaudus

adult

Gulls - *Laridae*

The family of Gulls is found worldwide, most of the species are migratory birds. The body length ranges from 20 cm to 75 cm. Gulls are strongly built, they have long, pointed wings and a rather long tail. Their feet are webbed. They are very good fliers who often sail or glide. They can also swim well, but few species dive. They often breed in large colonies.

Bonaparte's Gull
de: Bonapartemöwe
fr: Mouette de Bonaparte
es: Gaviota de Bonaparte
ja: ボナパルトカモメ
cn: 博氏鸥
Chroicocephalus philadelphia

breeding

Laughing Gull
de: Aztekenmöwe
fr: Mouette atricille
es: Gaviota Guanaguanare
ja: ワライカモメ
cn: 笑鸥
Leucophaeus atricilla

breeding

Ring-billed Gull
   de: Ringschnabelmöwe
   fr: Goéland à bec cerclé
   es: Gaviota de Delaware
   ja: クロワカモメ
   cn: 环嘴鸥
Larus delawarensis

Herring Gull
   de: Silbermöwe
   fr: Goéland argenté
   es: Gaviota Argéntea
   ja: セグロカモメ
   cn: 银鸥
Larus argentatus

Great Black-backed Gull
   de: Mantelmöwe
   fr: Goéland marin
   es: Gavión Atlántico
   ja: オオカモメ
   cn: 大黑背鸥
Larus marinus

Brown Noddy
   de: Noddi
   fr: Noddi brun
   es: Tiñosa Boba
   ja: クロアジサシ
   cn: 白顶玄燕鸥
Anous stolidus

Black Noddy
 de: Weißkopfnoddi
 fr: Noddi noir
 es: Tiñosa Menuda
 ja: ヒメクロアジサシ
 cn: 玄燕鸥
Anous minutus

adult, pullus

Sooty Tern
 de: Rußseeschwalbe
 fr: Sterne fuligineuse
 es: Charrán Sombrío
 ja: セグロアジサシ
 cn: 乌燕鸥
Onychoprion fuscatus

adult, Egg(s)

Bridled Tern
 de: Zügelseeschwalbe
 fr: Sterne bridée
 es: Charrán Embridado
 ja: マミジロアジサシ
 cn: 褐翅燕鸥
Onychoprion anaethetus

adult

Least Tern
 de: Amerikanische Zwergseeschwalbe
 fr: Petite Sterne
 es: Charrancito Americano
 ja: アメリカコアジサシ
 cn: 小白额燕鸥
Sternula antillarum

adult

[AU] Gull-billed Tern
de: Lachseeschwalbe
fr: Sterne hansel
es: Pagaza Piconegra
ja: ハシブトアジサシ
cn: 鸥嘴噪鸥
Gelochelidon nilotica

[AU] Black Tern
de: Trauerseeschwalbe
fr: Guifette noire
es: Fumarel Común
ja: ハシグロクロハラアジサシ
cn: 黑浮鸥
Chlidonias niger

[PD] Roseate Tern
de: Rosenseeschwalbe
fr: Sterne de Dougall
es: Charrán Rosado
ja: ベニアジサシ
cn: 粉红燕鸥
Sterna dougallii

[AU] Common Tern
de: Flussseeschwalbe
fr: Sterne pierregarin
es: Charrán Común
ja: アジサシ
cn: 普通燕鸥
Sterna hirundo

Arctic Tern
- de: Küstenseeschwalbe
- fr: Sterne arctique
- es: Charrán Artico
- ja: キョクアジサシ
- cn: 北极燕鸥

Sterna paradisaea

Forster's Tern
- de: Forsterseeschwalbe
- fr: Sterne de Forster
- es: Charrán de Forster
- ja: メリケンアジサシ
- cn: 弗氏燕鸥

Sterna forsteri

Royal Tern
- de: Königsseeschwalbe
- fr: Sterne royale
- es: Charrán Real
- ja: アメリカオオアジサシ
- cn: 橙嘴凤头燕鸥

Thalasseus maximus

Sandwich Tern
- de: Brandseeschwalbe
- fr: Sterne caugek
- es: Charrán Patinegro
- ja: サンドイッチアジサシ
- cn: 白嘴端凤头燕鸥

Thalasseus sandvicensis

www.avitopia.net/bird.en/?vid=1602203

Black Skimmer
- de: Schwarzmantel-Scherenschnabel
- fr: Bec-en-ciseaux noir
- es: Rayador Americano
- ja: クロハサミアジサシ
- cn: 黑剪嘴鸥

Rynchops niger

Pigeons and Doves - *Columbidae*

The family of Pigeons and Doves is found all over the world except in the coldest regions. The body lengths range from 15 cm to 84 cm. They have medium-sized wings and often a long tail. The beak is rather short and not strong. The sexes are mostly the same. Their diet is predominantly vegetarian. The naked young birds are fed 'pigeon milk', a secretion that is formed in the parents' goiter.

Common Pigeon
- de: Felsentaube
- fr: Pigeon biset
- es: Paloma Bravía
- ja: カワラバト(ドバト)
- cn: 原鸽

Columba livia

Introduced.

www.avitopia.net/bird.en/?vid=1650101

Scaly-naped Pigeon
- de: Antillentaube
- fr: Pigeon à cou rouge
- es: Paloma Isleña
- ja: アカエリバト
- cn: 鳞枕鸽

Patagioenas squamosa

White-crowned Pigeon
de: Weißscheiteltaube
fr: Pigeon à couronne blanche
es: Paloma Coronita
ja: シロボウシバト
cn: 白顶鸽
Patagioenas leucocephala
Near threatened.

Plain Pigeon
de: Rosenschultertaube
fr: Pigeon simple
es: Paloma Boba
ja: アンチルバト
cn: 纯色鸽
Patagioenas inornata
Near threatened.

Ring-tailed Pigeon
de: Karibentaube
fr: Pigeon de la Jamaïque
es: Paloma Jamaicana
ja: ジャマイカオビオバト
cn: 环尾鸽
Patagioenas caribaea
Endemic.
Vulnerable.

Eurasian Collared Dove
de: Türkentaube
fr: Tourterelle turque
es: Tórtola Turca
ja: シラコバト
cn: 灰斑鸠
Streptopelia decaocto
Introduced.

www.avitopia.net/bird.en/?kom=1650507
www.avitopia.net/bird.en/?vid=1650507

Common Ground Dove
de: Sperlingstäubchen
fr: Colombe à queue noire
es: Columbina Común
ja: スズメバト
cn: 地鳩
Columbina passerina

ssp. jamaicensis adult

Blue-headed Quail-Dove
de: Kubataube
fr: Colombe à tête bleue
es: Paloma-perdiz Cubana
ja: クロヒゲバト
cn: 蓝头鹌鸠
Starnoenas cyanocephala
Introduced.
Endangered.

adult

Crested Quail-Dove
de: Kurzschopftaube
fr: Colombe versicolore
es: Paloma-perdiz Jamaicana
ja: カンムリウズラバト
cn: 凤头鹌鸠
Geotrygon versicolor
Endemic.
Near threatened.

adult

Ruddy Quail-Dove
de: Bergtaube
fr: Colombe rouviolette
es: Paloma-perdiz Común
ja: ヤマウズラバト
cn: 红鹑鸠
Geotrygon montana

adult

Caribbean Dove
   de: Jamaikataube
   fr: Colombe de la Jamaïque
   es: Paloma Montaraz Jamaicana
   ja: シロハラシャコバト
   cn: 白腹棕翅鸠
   Leptotila jamaicensis

ssp. jamaicensis adult

White-winged Dove
   de: Weißflügeltaube
   fr: Tourterelle à ailes blanches
   es: Zenaida Aliblanca
   ja: ハジロバト
   cn: 白翅哀鸽
   Zenaida asiatica

adult

Zenaida Dove
   de: Liebestaube
   fr: Tourterelle à queue carrée
   es: Zenaida Caribeña
   ja: シマハジロバト
   cn: 鸣哀鸽
   Zenaida aurita

adult

Mourning Dove
   de: Carolinataube
   fr: Tourterelle triste
   es: Zenaida Huilota
   ja: ナゲキバト
   cn: 哀鸽
   Zenaida macroura

adult

Cuckoos - *Cuculidae*

The family of Cuckoos is found worldwide except in the coldest areas. Many species are migratory birds. Cuckoos often have slender bodies and very long tails. The body length ranges from 17 cm to 70 cm. Except for the Roadrunners, the legs are short, with two toes pointing forward and two pointing backwards. The sexes are mostly the same. They feed mainly on insects. Many species are pronounced brood parasites. After hatching, the young cuckoos regularly push the other nest siblings out of the nest. The non-parasitic species are able to build nests. With the exception of the Anis, cuckoos are loners.

Smooth-billed Ani
de: Glattschnabel-Ani
fr: Ani à bec lisse
es: Garrapatero Aní
ja: オオハシカッコウ
cn: 滑嘴犀鵑
Crotophaga ani

Yellow-billed Cuckoo
de: Gelbschnabelkuckuck
fr: Coulicou à bec jaune
es: Cuclillo Piquigualdo
ja: キバシカッコウ
cn: 黄嘴美洲鹃
Coccyzus americanus

Mangrove Cuckoo
de: Mangrovekuckuck
fr: Coulicou manioc
es: Cuclillo de Manglar
ja: マングロアブカッコウ
cn: 红树美洲鹃
Coccyzus minor

Black-billed Cuckoo
- de: Schwarzschnabelkuckuck
- fr: Coulicou à bec noir
- es: Cuclillo Piquinegro
- ja: ハシグロカッコウ
- cn: 黑嘴美洲鹃

Coccyzus erythropthalmus

Chestnut-bellied Cuckoo
- de: Regenkuckuck
- fr: Piaye de pluie
- es: Cuco Picogordo de Jamaica
- ja: ハイムネリスカッコウ
- cn: 栗腹鹃

Coccyzus pluvialis
Endemic.

Jamaican Lizard Cuckoo
- de: Jamaikakuckuck
- fr: Tacco de la Jamaïque
- es: Cuco-lagartero Jamaicano
- ja: トゲカッコウ
- cn: 牙买加蜥鹃

Coccyzus vetula
Endemic.

Barn owls - *Tytonidae*

The family of Barn Owls includes only a few species, but one of them is a true cosmopolitan. It occurs on all continents and only avoids the colder areas of the earth. The body length ranges from 23 cm to 55 cm. The legs are quite long, the central claw is comb-like. The head is endowed with a conspicuous veil and has a curved beak. Barn owls are nocturnal hunters that fly noiselessly near the ground. They can hear directionally and are able to locate their prey by hearing only.

Barn Owl
de: Schleiereule
fr: Effraie des clochers
es: Lechuza Común
ja: メンフクロウ
cn: 仓鸮
Tyto alba

Owls - *Strigidae*

The family of Owls is found worldwide. They have compact bodies (13 cm - 70 cm) and usually wide wings and rounded tails. The toes are strong, one of which is a turning toe that helps with grip. The head is large, the neck short. The eyes are directed rather forward, the beak is short and hook-shaped. Owls mainly hunt at night, benefiting from their noiseless flight and sharp hearing.

Burrowing Owl
de: Kaninchenkauz
fr: Chevêche des terriers
es: Mochuelo de Madriguera
ja: アナホリフクロウ
cn: 穴小鸮
Athene cunicularia

Stygian Owl
   de: Styxeule
   fr: Hibou maître-bois
   es: Búho Negruzco
   ja: ナンベイトラフズク
   cn: 乌耳鸮
Asio stygius

adult

Short-eared Owl
   de: Sumpfohreule
   fr: Hibou des marais
   es: Búho Campestre
   ja: コミミズク
   cn: 短耳鸮
Asio flammeus

adult

Jamaican Owl
   de: Jamaikaeule
   fr: Hibou de la Jamaïque
   es: Búho Jamaicano
   ja: ジャマイカズク
   cn: 牙买加鸮
Pseudoscops grammicus
Endemic.

adult

Nightjars - *Caprimulgidae*

The family of Nightjars is found all over the world, with the exception of the colder areas and New Zealand. The body length is between 19 cm and 30 cm, but elongated wing feathers appear in two species. Nightjars have long wings and tails, but their legs, toes, and claws are small. The beak is also small, but the throat is very wide. They are crepuscular and nocturnal and sit motionless on the ground or on a branch during the day. They feed on insects that they prey on in flight. The eggs are usually laid directly on the ground. There are also species among the nightjars that hibernate.

Common Nighthawk
de: Nachtfalke
fr: Engoulevent d'Amérique
es: Añapero Yanqui
ja: アメリカヨタカ
cn: 美洲夜鷹
Chordeiles minor

Antillean Nighthawk
de: Antillennachtschwalbe
fr: Engoulevent piramidig
es: Añapero Querequeté
ja: カリブヨタカ
cn: 安島夜鷹
Chordeiles gundlachii

Jamaican Pauraque
de: Jamaikanachtschwalbe
fr: Engoulevent de la Jamaïque
es: Chotacabras Jamaicano
ja: ジャマイカヨタカ
cn: 牙买加夜鷹
Siphonorhis americana †
Endemic.
Last seen 1859.

Eastern Whip-poor-will
 de: Schwarzkehl-Nachtschwalbe
 fr: Engoulevent bois-pourri
 es: Chotacabras Cuerporruín
 ja: ホイップアアウイルヨタカ
 cn: 三声夜鹰
Antrostomus vociferus

adult

Potoos - *Nyctibiidae*

The family of Potoos occurs from Mexico south to Argentina and on some islands in the Caribbean. They become 40 cm to 50 cm long. The body, wings and tail are long. The eyes are very large, the beak is short, narrow and has a curved tip. The feet are very small. The Potoos live in forests and are nocturnal loners. During the day they sit upright motionless on a branch. They mainly eat insects, which they catch from a perch on short flights. The only egg is laid in a cavity in a tree and incubated by both parents. The young are altricial.

Northern Potoo
 de: Mexikotagschläfer
 fr: Ibijau jamaïcain
 es: Nictibio Jamaicano
 ja: ハイイロタチヨタカ
 cn: 北林鸱
Nyctibius jamaicensis

ssp. jamaicensis[8] adult

Swifts - *Apodidae*

The family of Swifts is globally distributed except in the coldest regions. They are small birds, 9 cm to 23 cm long. The wings are long and pointed, the legs and feet are very small and the beak is small with a crooked point and a wide throat. Sails are perfectly adapted to life in the air, and some species are able to spend the night in flight and to mate. They are excellent and fast fliers, on the other hand, many species cannot take off from the ground. The nests of a few species are made entirely of saliva and are considered a delicacy in Chinese cuisine. Some species of salangan have the exceptional echolocation capability. They use this to orient themselves in underground cave systems, where their nesting sites are.

American Black Swift
de: Schwarzsegler
fr: Martinet sombre
es: Vencejo Negro
ja: クロムジアマツバメ
cn: 黑雨燕
Cypseloides niger

White-collared Swift
de: Halsbandsegler
fr: Martinet à collier blanc
es: Vencejo Acollarado
ja: クビワアマツバメ
cn: 白领黑雨燕
Streptoprocne zonaris

Chimney Swift
de: Schornsteinsegler
fr: Martinet ramoneur
es: Vencejo de Chimenea
ja: エントツアマツバメ
cn: 烟囱雨燕
Chaetura pelagica
Near threatened.

Antillean Palm Swift
   de: Kubasegler
   fr: Martinet petit-rollé
   es: Vencejillo Antillano
   ja: アメリカヤシアマツバメ
   cn: 西印棕雨燕
Tachornis phoenicobia

adult

Hummingbirds - *Trochilidae*

The family of Hummingbirds occurs in South and Central America and, with fewer species, also in North America. The body size ranges from 6 cm to 33 cm, hence the smallest birds on earth are among the hummingbirds. They have long narrow wings, short legs and small weak feet. The plumage is very differently colored and iridescent colors often appear. Their habitat is mainly in the forest. Their flying dexterity is incredible, they can not only fly in place, but also backwards. They eat insects, spiders and nectar. Their high energy consumption demands a large amount of food. The nest is a deep bowl made of moss, down and cobwebs with two eggs that are hatched by the female.

Jamaican Mango
   de: Jamaikamango
   fr: Mango de la Jamaïque
   es: Mango Jamaicano
   ja: ジャマイカマンゴアハチドリ
   cn: 牙买加芒果蜂鸟
Anthracothorax mango
Endemic.

www.avitopia.net/bird.en/?vid=1976907

adult

Ruby-throated Hummingbird
   de: Rubinkehlkolibri
   fr: Colibri à gorge rubis
   es: Colibrí Gorgirrubí
   ja: ノドアカハチドリ
   cn: 红喉北蜂鸟
Archilochus colubris

♂ adult

ssp. minima(e) ♂ adult

Photo W.J.Daunicht

AU Vervain Hummingbird
de: Zwergelfe
fr: Colibri nain
es: Colibrí Zumbadorcito
ja: コビトハチドリ
cn: 小吸蜜蜂鸟
Mellisuga minima

♂ adult

Photo W.J.Daunicht

AU Streamertail
de: Wimpelschwanz
fr: Colibri à tête noire
es: Colibrí Portacintas
ja: フキナガシハチドリ
cn: 红嘴长尾蜂鸟
Trochilus polytmus
Endemic.

Todis - *Todidae*

The family of Todies contains only 5 species, all of which live in the Greater Antilles or the adjacent small islands. The tiny birds range in length from 10 cm to 11.5 cm. They have long flattened beaks with serrated edges. Their preferred habitat is the undergrowth in the forest. They eat small invertebrates, but also small lizards, which they hunt from a perch. The nest lies in a chamber at the end of a tunnel up to 60 cm long, which they dig up a steep banks or into a rotten tree trunk. Both parents take care of the altricial nestlings, which they feed up to 140 times a day, the highest rate among birds.

adult

Photo W.J.Daunicht

AU Jamaican Tody
de: Grüntodi
fr: Todier de la Jamaïque
es: Barrancolí Jamaicano
ja: ジャマイカコビトドリ
cn: 短尾鴗
Todus todus
Endemic.

Kingfishers - *Alcedinidae*

The family of Kingfishers are found worldwide except in the coldest areas and some islands. They are between 10 and 45 cm long. Kingfishers have a stocky body, short wings, and tiny to very long tails. The head is large, the neck short, and the beak long and thick. When hunting, they come down from a viewing point. Some species are capable of hovering flight. They breed in caves.

Belted Kingfisher
  de: Gürtelfischer
  fr: Martin-pêcheur d'Amérique
  es: Martín Gigante Norteamericano
  ja: アメリカヤマセミ
  cn: 白腹鱼狗
Megaceryle alcyon

adult

Woodpeckers - *Picidae*

The family of Woodpeckers is found worldwide, but not in Madagascar, Australia and most of the Indonesian islands. The body length ranges from 9 cm to 60 cm. The feet have 3 or 4 toes, two of which point forward. They have a chunky head and a straight, usually powerful beak. The sexes are mostly different. Woodpeckers mostly inhabit trees. Most species feed on insects, which they pull out with their long and flexible tongue. When climbing vertical logs, the stiff tail serves as a support. They make a cave in a tree to breed.

Jamaican Woodpecker
  de: Jamaikaspecht
  fr: Pic de la Jamaïque
  es: Carpintero Jamaicano
  ja: ジャマイカシマセゲラ
  cn: 牙买加啄木鸟
Melanerpes radiolatus
Endemic.

♂ adult

Yellow-bellied Sapsucker
- de: Gelbbauch-Saftlecker
- fr: Pic maculé
- es: Chupasavia Norteño
- ja: シルスイキツツキ
- cn: 黄腹吸汁啄木鸟

Sphyrapicus varius

♂ adult

Falcons - *Falconidae*

The family of Falcons is found on every continent except Antarctica. Their length ranges from 15 cm to 65 cm. Hawks have long, pointed wings, a half-length tail and short legs that end in long toes with curved claws. The beak is short and usually has a so-called 'falcon tooth' in the upper beak. The flight is determined and fast. Some species strike their prey in flight after a chase, while other species take them on the ground after diving. In fact, the fastest fliers among birds belong to this family.

Northern Crested Caracara
- de: Nordkarakara
- fr: Caracara du Nord
- es: Carancho Norteño
- ja: カンムリカラカラ
- cn: 巨隼

Caracara cheriway

adult

American Kestrel
- de: Buntfalke
- fr: Crécerelle d'Amérique
- es: Cernícalo Americano
- ja: アメリカチョウゲンボウ
- cn: 美洲隼

Falco sparverius

♂ adult

Merlin
  de: Merlin
  fr: Faucon émerillon
  es: Esmerejón
  ja: コチョウゲンボウ
  cn: 灰背隼
Falco columbarius

♂ adult

Peregrine Falcon
  de: Wanderfalke
  fr: Faucon pèlerin
  es: Halcón Peregrino
  ja: ハヤブサ
  cn: 游隼
Falco peregrinus

adult

New World and African Parrots - *Psittacidae*

This family is limited to the continents of America and Africa. Most of the species in this family live in America, with only a few species in Africa. The size ranges from 12 cm to 100 cm. Many species have bright colors, caused by the Dyck textures in the feathers, in which incident light is refracted. Most of the species are cavity breeders.

Black-billed Amazon
  de: Rotspiegelamazone
  fr: Amazone verte
  es: Amazona Jamaicana Piquioscura
  ja: ハシグロボウシインコ
  cn: 黑嘴鹦哥
Amazona agilis
Endemic.
Vulnerable.

adult

Yellow-billed Amazon
de: Jamaikaamazone
fr: Amazone sasabé
es: Amazona Jamaicana Piquiclara
ja: エリマキボウシインコ
cn: 黄嘴鹦哥
Amazona collaria
Endemic.
Vulnerable.

Green-rumped Parrotlet
de: Grünbürzel-Sperlingspapagei
fr: Toui été
es: Cotorrita Culiverde
ja: テリルリハシインコ
cn: 绿腰鹦哥
Forpus passerinus
Introduced.

Jamaican Parakeet
de: Kleiner Aztekensittich
fr: Conure naine
es: Aratinga Pechisucia
ja: チャムネメキシコインコ
cn: 绿喉鹦哥
Eupsittula nana
Near threatened.

Tyrant-flycatchers - *Tyrannidae*

The family of Tyrants is common across America. Their length ranges from 8 cm to 40 cm. This family is one of the most diverse in the bird world. Many species hunt from a perch. They defend their territory very fiercely and attack even larger birds. They have very different nest shapes. 2 to 6 eggs are laid, which are usually hatched by the female. The altricial chicks are looked after by both adult birds.

Jamaican Elaenia
   de: Cottaelaenie
   fr: Élénie de la Jamaïque
   es: Fiofío Jamaicano
   ja: ジャマイカキクイタダキモドキ
   cn: 才头加伊拉鹟
Myiopagis cotta
Endemic.

Greater Antillean Elaenia
   de: Antillenelaenie
   fr: Élénie sara
   es: Fiofío Canoso
   ja: アンチルシラギクタイランチョウ
   cn: 安岛拟霸鹟
Elaenia fallax

Eastern Wood Pewee
   de: Östlicher Waldtyrann
   fr: Pioui de l'Est
   es: Pibí Oriental
   ja: モリタイランチョウ
   cn: 东绿霸鹟
Contopus virens

Jamaican Pewee
- de: Jamaikaschnäppertyrann
- fr: Moucherolle de la Jamaïque
- es: Pibí Jamaicano
- ja: ジャマイカヒタキモドキ
- cn: 牙买加绿霸鹟

Contopus pallidus
Endemic.

Acadian Flycatcher
- de: Buchentyrann
- fr: Moucherolle vert
- es: Mosquero Verdoso
- ja: ミドリメジロハエトリ
- cn: 绿纹霸鹟

Empidonax virescens

Sad Flycatcher
- de: Jamaikatyrann
- fr: Tyran triste
- es: Copetón Jamaicano
- ja: ジャマイカヒタキモドキ
- cn: 牙买加蝇霸鹟

Myiarchus barbirostris
Endemic.

Great Crested Flycatcher
- de: Schnäppertyrann
- fr: Tyran huppé
- es: Copetón Viajero
- ja: オオヒタキモドキ
- cn: 大冠蝇霸鹟

Myiarchus crinitus

Rufous-tailed Flycatcher
 de: Rotschwanztyrann
 fr: Tyran à queue rousse
 es: Copetón Colirrufo
 ja: アカオオオヒタキモドキ
 cn: 棕尾蝇霸鹟
Myiarchus validus
Endemic.

Stolid Flycatcher
 de: Dickkopftyrann
 fr: Tyran grosse-tête
 es: Copetón Bobito
 ja: アカオヒタキモドキ
 cn: 憨蝇霸鹟
Myiarchus stolidus

Eastern Kingbird
 de: Königstyrann
 fr: Tyran tritri
 es: Tirano Oriental
 ja: オウサマタイランチョウ
 cn: 东王霸鹟
Tyrannus tyrannus

Grey Kingbird
 de: Grautyrann
 fr: Tyran gris
 es: Tirano Dominicano
 ja: ハイイロタイランチョウ
 cn: 灰王霸鹟
Tyrannus dominicensis

Loggerhead Kingbird
de: Bahamatyrann
fr: Tyran tête-police
es: Tirano Guatíbere
ja: オジロハイイロタイランチョウ
cn: 圆头王霸鹟
Tyrannus caudifasciatus

ssp. jamaicensis® adult

Tityras and allies - *Tityridae*

The family of Tityras and allies is found in the tropics of America. The species of this family have been distributed over the families of Tyrant-flycatchers, Manakins, and Cotingas until several DNA studies have been carried out. The body sizes range from 9.5 cm to 24 cm. Most species have large heads and short tails. They live in forests and other tree-lined landscapes.

Jamaican Becard
de: Jamaikabekarde
fr: Bécarde de la Jamaïque
es: Anambé Jamaicano
ja: クロカザリドリモドキ
cn: 牙买加厚嘴霸鹟
Pachyramphus niger
Endemic.

♂ adult

Vireos and Shrike-Babblers - *Vireonidae*

The family of the Vireos and Shrike-babblers was composed only recently on the basis of DNA studies from the Vireos of America and two genera from the Oriental region. Since they are native to the warm regions, they do not migrate far. They are small birds between 10 cm and 20 cm in length. They live mainly on insects, but also on fruits. They rarely look for food on the ground. The bowl-shaped nest is built hanging in a horizontal fork of a branch.

Blue Mountain Vireo
- de: Osburnvireo
- fr: Viréo d'Osburn
- es: Vireo de las Monatñas Azules
- ja: ブルアマウンテンモズモドキ
- cn: 山莺雀

Vireo osburni
Endemic.
Near threatened.

Jamaican Vireo
- de: Jamaikavireo
- fr: Viréo de la Jamaïque
- es: Vireo Jamaiquino
- ja: ジャマイカモズモドキ
- cn: 牙买加莺雀

Vireo modestus
Endemic.

Yellow-throated Vireo
- de: Gelbkehlvireo
- fr: Viréo à gorge jaune
- es: Vireo de Cuello Amarilla
- ja: キノドモズモドキ
- cn: 黄喉莺雀

Vireo flavifrons

Red-eyed Vireo
de: Rotaugenvireo
fr: Viréo aux yeux rouges
es: Vireo de Ojos Rojos
ja: アカメモズモドキ
cn: 红眼莺雀
Vireo olivaceus

Black-whiskered Vireo
de: Bartvireo
fr: Viréo à moustaches
es: Vireo de Bigotes Negros
ja: ホオヒゲモズモドキ
cn: 黑髭莺雀
Vireo altiloquus

Ravens - *Corvidae*

The family of Ravens occurs worldwide with the exception of New Zealand and some islands. The body length is between 18 cm and 70 cm; so among them are the largest of all songbirds. Ravens have powerful bills and often hold the food with their feet when eating. They are curious and one of the most intelligent species in the entire bird world.

Jamaican Crow
de: Jamaikakrähe
fr: Corneille de la Jamaïque
es: Cuervo Jamaiquino
ja: ジャマイカガラス
cn: 牙买加乌鸦
Corvus jamaicensis
Endemic.

Swallows - *Hirundinidae*

The family of Swallows is found all over the world with the exception of the coldest regions, many species are migratory birds. They are quite small birds with a body length of 10 cm to 23 cm. The wings are long and pointed, the legs and feet small, the beak short with a wide throat. Swallows are fast and extremely agile fliers. They feed exclusively on flying insects. They nest in dug earth caves, natural caves or bowl-shaped mud nests. They breed up to three times a year.

Northern Rough-winged Swallow
  de: Nördliche Rauflügelschwalbe
  fr: Hirondelle à ailes hérissées
  es: Golondrina Aserrada
  ja: キタオビナシショウドウツバメ
  cn: 中北美毛翅燕
Stelgidopteryx serripennis

Purple Martin
  de: Purpurschwalbe
  fr: Hirondelle noire
  es: Golondrina Purpúrea
  ja: ムラサキツバメ
  cn: 紫崖燕
Progne subis

Caribbean Martin
  de: Dominikanerschwalbe
  fr: Hirondelle à ventre blanc
  es: Golondrina Caribeña
  ja: シロハラムラサキツバメ
  cn: 加勒比崖燕
Progne dominicensis

Tree Swallow
- de: Sumpfschwalbe
- fr: Hirondelle bicolore
- es: Golondrina Bicolor
- ja: ミドリツバメ
- cn: 双色树燕

Tachycineta bicolor

Golden Swallow
- de: Antillenschwalbe
- fr: Hirondelle dorée
- es: Golondrina Dorada
- ja: キンイロツバメ
- cn: 金色树燕

Tachycineta euchrysea
Vulnerable.

Sand Martin
- de: Uferschwalbe
- fr: Hirondelle de rivage
- es: Avión Zapador
- ja: ショウドウツバメ
- cn: 崖沙燕

Riparia riparia

www.avitopia.net/bird.en/?vid=4450904
www.avitopia.net/bird.en/?aud=4450904

Barn Swallow
- de: Rauchschwalbe
- fr: Hirondelle rustique
- es: Golondrina Común
- ja: ツバメ
- cn: 家燕

Hirundo rustica

www.avitopia.net/bird.en/?vid=4451201

American Cliff Swallow
  de: Fahlstirnschwalbe
  fr: Hirondelle à front blanc
  es: Golondrina Risquera
  ja: サンショクツバメ
  cn: 美洲燕
Petrochelidon pyrrhonota

Cave Swallow
  de: Höhlenschwalbe
  fr: Hirondelle à front brun
  es: Golondrina Pueblera
  ja: セスジツバメ
  cn: 穴崖燕
Petrochelidon fulva

Gnatcatchers - *Polioptilidae*

The family of Gnatcatchers occurs in America from southern Canada to Argentina. They are 10 cm to 13 cm long, slender long-tailed birds that are related to the wrens. Their diet is mainly insects. Both parents together build a bowl-shaped nest made of fine plant material, hair and cobwebs in trees up to 24 m above the ground. Both partners incubate a clutch of 3 to 5 eggs for 13 to 17 days.

Blue-grey Gnatcatcher
  de: Blaumückenfänger
  fr: Gobemoucheron gris-bleu
  es: Perlita Común
  ja: ブユムシクイ
  cn: 灰蓝蚋莺
Polioptila caerulea

Thrushes - *Turdidae*

The family of Thrushes is distributed worldwide and is even found on many small islands in the Pacific, only missing in Antarctica and New Zealand. But Blackbirds and Song thrushes have been introduced there and have reproduced so much that they are now among the most common birds. Thrushes mainly feed on insects and other invertebrates, but berries also play a role in winter.

Rufous-throated Solitaire
de: Bartklarino
fr: Solitaire siffleur
es: Clarín de Garganta Rufa
ja: アカノドヒトリツグミ
cn: 棕喉孤鸫
Myadestes genibarbis

Veery
de: Wilsondrossel
fr: Grive fauve
es: Tordo Cachetón
ja: ビリアチャツグミ
cn: 棕夜鸫
Catharus fuscescens

Grey-cheeked Thrush
de: Grauwangendrossel
fr: Grive à joues grises
es: Tordo de Cara Gris
ja: ハイイロチャツグミ
cn: 灰颊夜鸫
Catharus minimus

Swainson's Thrush
  de: Zwergdrossel
  fr: Grive à dos olive
  es: Tordo Olivo
  ja: オリアブチャツグミ
  cn: 斯氏夜鸫
Catharus ustulatus

Wood Thrush
  de: Walddrossel
  fr: Grive des bois
  es: Zorzal Maculado
  ja: モリツグミ
  cn: 棕林鸫
Hylocichla mustelina
Near threatened.

White-eyed Thrush
  de: Weißaugendrossel
  fr: Merle aux yeux blancs
  es: Tordo de Ojos Blancos
  ja: メジロムジツグミ
  cn: 白眼鸫
Turdus jamaicensis
Endemic.

White-chinned Thrush
  de: Weißkinndrossel
  fr: Merle à miroir
  es: Zorzal de Barbilla Blanca
  ja: アゴジロツグミ
  cn: 白颏鸫
Turdus aurantius
Endemic.

Mockingbirds - *Mimidae*

The family of Mockingbirds is distributed in America from southern Canada over the Caribbean and the Galapagos Islands to Argentina. The body length is 20 cm to 30 cm. They have short wings and a long tail. Their legs are quite long. They have a large repertoire of vocalizations that can include more than a thousand song phrases. In addition, some species can imitate the voices of other birds and also technical sounds, e.g. car alarms. Most species do not fly much. The nest is a large bowl that is built in a bush or on the ground. The altricial young are looked after by both adult birds.

Grey Catbird
de: Katzenvogel
fr: Moqueur chat
es: Sinsonte Maullador
ja: ネコマネドリ
cn: 灰嘲鸫
Dumetella carolinensis

Bahama Mockingbird
de: Gundlachspottdrossel
fr: Moqueur des Bahamas
es: Sinsonte de las Bahamas
ja: バハママネシツグミ
cn: 巴哈马小嘲鸫
Mimus gundlachii

Northern Mockingbird
de: Spottdrossel
fr: Moqueur polyglotte
es: Sinsonte Común
ja: マネシツグミ
cn: 小嘲鸫
Mimus polyglottos

Starlings - *Sturnidae*

The family of Strlings was originally only distributed in the Old World, but the common star was introduced in America and is now widespread there. The body length ranges from 18 cm to 43 cm. Many species have iridescent plumage. The tail is usually short, more rarely long. Unlike thrushes, starlings do not hop, but run with alternating steps. They fly well and the formation flights of large flocks of starlings are impressive. Most species breed in tree hollows, but other nesting techniques also occur, including large community nests. Starlings are omnivores, one reason for their assertiveness as colonists.

Common Starling
de: Star
fr: Étourneau sansonnet
es: Estornino Pinto
ja: ホシムクドリ
cn: 紫翅椋鸟
Sturnus vulgaris

www.avitopia.net/bird.en/?kom=5326201

♂ breeding

Common Myna
de: Hirtenmaina
fr: Martin triste
es: Mainá Común
ja: カバイロハッカ
cn: 家八哥
Acridotheres tristis
Introduced.

adult

Wagtails and Pipits - *Motacillidae*

The family of Wagtails is found worldwide except in the coldest areas, many species are migratory birds. They are slender birds with a body length of 13 cm to 22 cm. All species are ground birds, but they can also fly well. First and foremost, they are insectivores.

Buff-bellied Pipit
de: Pazifischer Wasserpieper
fr: Pipit d'Amérique
es: Bisbita Norteamericano
ja: アメリカタヒバリ
cn: 黄腹鹨
Anthus rubescens

Waxwings - *Bombycillidae*

The family of Waxwings is native to the northern subarctic hemisphere. They are about 18 cm long. They are sociable tree dwellers who mainly feed on berries. In some years they invade areas that are outside the normal wintering range.

Cedar Waxwing
de: Zedernseidenschwanz
fr: Jaseur d'Amérique
es: Ampelis Americano
ja: ヒメレンジャク
cn: 雪松太平鸟
Bombycilla cedrorum

American Warblers - *Parulidae*

The family of American Warblers is restricted to America. However, they are long-distance migratory birds and can therefore also be found as random visitors in distant places. The birds are relatively small at 11 cm to 18 cm. All of them have only nine hand wing feathers. They feed almost exclusively on insects.

Ovenbird
- de: Pieperwaldsänger
- fr: Paruline couronnée
- es: Chipe de Tierra
- ja: カマドムシクイ
- cn: 橙顶灶莺

Seiurus aurocapilla

adult

Worm-eating Warbler
- de: Haldenwaldsänger
- fr: Paruline vermivore
- es: Chipe Gusanero
- ja: フタスジアメリカムシクイ
- cn: 食虫莺

Helmitheros vermivorum

adult

Louisiana Waterthrush
- de: Stelzenwaldsänger
- fr: Paruline hochequeue
- es: Chipe de Agua Sureño
- ja: ミナミミズツグミ
- cn: 白眉灶莺

Parkesia motacilla

adult

Northern Waterthrush
de: Drosselwaldsänger
fr: Paruline des ruisseaux
es: Chipe de Agua Norteño
ja: キタミズツグミ
cn: 黄眉灶莺
Parkesia noveboracensis

Golden-winged Warbler
de: Goldflügel-Waldsänger
fr: Paruline à ailes dorées
es: Chipe de Alas Doradas
ja: キンバネアメリカムシクイ
cn: 金翅虫森莺
Vermivora chrysoptera
Near threatened.

Blue-winged Warbler
de: Blauflügel-Waldsänger
fr: Paruline à ailes bleues
es: Chipe de Alas Azules
ja: アオバネアメリカムシクイ
cn: 蓝翅虫森莺
Vermivora cyanoptera

Black-and-white Warbler
de: Kletterwaldsänger
fr: Paruline noir et blanc
es: Chipe Rayado
ja: シロクロアメリカムシクイ
cn: 黑白森莺
Mniotilta varia

Prothonotary Warbler
  de: Zitronenwaldsänger
  fr: Paruline orangée
  es: Chipe Anaranjado
  ja: オウゴンアメリカムシクイ
  cn: 蓝翅黄森莺
  Protonotaria citrea

♂ breeding

Swainson's Warbler
  de: Swainsonwaldsänger
  fr: Paruline de Swainson
  es: Chipe de Swainson
  ja: チャカブリアメリカムシクイ
  cn: 白眉食虫莺
  Limnothlypis swainsonii

adult

Tennessee Warbler
  de: Brauenwaldsänger
  fr: Paruline obscure
  es: Chipe Peregrino
  ja: マミジロアメリカムシクイ
  cn: 灰冠虫森莺
  Oreothlypis peregrina

adult

Connecticut Warbler
  de: Augenring-Waldsänger
  fr: Paruline à gorge grise
  es: Chipe de Connecticut
  ja: ハイムネアメリカムシクイ
  cn: 灰喉地莺
  Oporornis agilis

adult

Mourning Warbler
de: Graukopf-Waldsänger
fr: Paruline triste
es: Chipe Enlutado
ja: ノドグロアメリカムシクイ
cn: 黑胸地莺
Geothlypis philadelphia

Kentucky Warbler
de: Kentuckywaldsänger
fr: Paruline du Kentucky
es: Chipe de Mejillas Negras
ja: メガネアメリカムシクイ
cn: 黄腹地莺
Geothlypis formosa

Common Yellowthroat
de: Weidengelbkehlchen
fr: Paruline masquée
es: Chipe de Cara Negra
ja: カオグロアメリカムシクイ
cn: 黄喉地莺
Geothlypis trichas

Arrowhead Warbler
de: Strichelwaldsänger
fr: Paruline de la Jamaïque
es: Chipe Cabeza de Flecha
ja: ヤジリアメリカムシクイ
cn: 尖头林莺
Setophaga pharetra
Endemic.

Hooded Warbler
- de: Kapuzenwaldsänger
- fr: Paruline à capuchon
- es: Chipe Careto
- ja: クロズキンアメリカムシクイ
- cn: 黑枕威森莺

Setophaga citrina

♂ adult

American Redstart
- de: Schnäpperwaldsänger
- fr: Paruline flamboyante
- es: Chipe Rey Americano
- ja: ハゴロモムシクイ
- cn: 橙尾鸲莺

Setophaga ruticilla

♂ adult

Cape May Warbler
- de: Tigerwaldsänger
- fr: Paruline tigrée
- es: Chipe Tigre
- ja: ホオアカアメリカムシクイ
- cn: 栗颊林莺

Setophaga tigrina

♂ breeding

Northern Parula
- de: Meisenwaldsänger
- fr: Paruline à collier
- es: Párula de Pecho Dorado
- ja: アサギアメリカムシクイ
- cn: 北森莺

Setophaga americana

♂ adult

Magnolia Warbler
de: Magnolienwaldsänger
fr: Paruline à tête cendrée
es: Chipe de Cola Fajeada
ja: シロオビアメリカムシクイ
cn: 纹胸林莺
Setophaga magnolia

non-breeding

Bay-breasted Warbler
de: Braunbrust-Waldsänger
fr: Paruline à poitrine baie
es: Chipe Castaño
ja: クリイロアメリカムシクイ
cn: 栗胸林莺
Setophaga castanea

♂ adult

American Yellow Warbler
de: Goldwaldsänger
fr: Paruline jaune
es: Chipe Amarillo
ja: キイロアメリカムシクイ
cn: 黄林莺
Setophaga petechia

adult

Chestnut-sided Warbler
de: Gelbscheitel-Waldsänger
fr: Paruline à flancs marron
es: Chipe Pardo-blanco
ja: ワキチャアメリカムシクイ
cn: 栗胁林莺
Setophaga pensylvanica

adult

Blackpoll Warbler
- de: Streifenwaldsänger
- fr: Paruline rayée
- es: Chipe Estriado
- ja: ズグロアメリカムシクイ
- cn: 白颊林莺

Setophaga striata

Black-throated Blue Warbler
- de: Blaurücken-Waldsänger
- fr: Paruline bleue
- es: Chipe Azul y Negro
- ja: ノドグロルリアメリカムシクイ
- cn: 黑喉蓝林莺

Setophaga caerulescens

♂ adult

Palm Warbler
- de: Palmenwaldsänger
- fr: Paruline à couronne rousse
- es: Chipe Palmero
- ja: ヤシアメリカムシクイ
- cn: 棕榈林莺

Setophaga palmarum

adult

Yellow-rumped Warbler
- de: Kronwaldsänger
- fr: Paruline à croupion jaune
- es: Chipe Coronado
- ja: キズタアメリカムシクイ
- cn: 黄腰林莺

Setophaga coronata

breeding

Yellow-throated Warbler
de: Goldkehl-Waldsänger
fr: Paruline à gorge jaune
es: Chipe de Garganta Amarilla
ja: キノドアメリカムシクイ
cn: 黄喉林莺

Setophaga dominica

adult

Prairie Warbler
de: Rostscheitel-Waldsänger
fr: Paruline des prés
es: Chipe Galano
ja: チャスジアメリカムシクイ
cn: 草原林莺

Setophaga discolor

adult

Black-throated Green Warbler
de: Grünwaldsänger
fr: Paruline à gorge noire
es: Chipe de Garganta Negra
ja: ノドグロミドリアメリカムシクイ
cn: 黑喉绿林莺

Setophaga virens

adult

Tanagers - *Thraupidae*

The family of Tanagers is the second largest of the passerine birds. It has an American distribution mainly in the tropics. Their body length ranges from 9 cm to 24 cm. Their plumage is usually brightly colored, but there are also black and white species. Tangerines are omnivores. The female builds the nest and incubates, but may be fed by the male. Both parents feed the young.

Saffron Finch
   de: Safranammer
   fr: Sicale bouton-d'or
   es: Semillero Basto
   ja: キンノジコ
   cn: 橙黄雀鹀
Sicalis flaveola
Introduced.

♂ adult

Bananaquit
   de: Zuckervogel
   fr: Sucrier à ventre jaune
   es: Platanera Común
   ja: マミジロミツドリ
   cn: 曲嘴森莺
Coereba flaveola

ssp. flaveola adult

Yellow-faced Grassquit
   de: Goldbraue
   fr: Sporophile grand-chanteur
   es: Tomeguín de la Tierra
   ja: キマユクビワスズメ
   cn: 黄脸草雀
Tiaris olivaceus

♂ adult

Black-faced Grasquit
de: Schwarzgesichtchen
fr: Sporophile cici
es: Semillerito Bicolor
ja: ニショクコメワリ
cn: 黑脸草雀
Tiaris bicolor

ssp. marchii ♂ adult

Orangequit
de: Braunlätzchen
fr: Pique-orange de la Jamaïque
es: Pinzón Jamaiquino
ja: ノドアカミツドリ
cn: 橙喉雀
Euneornis campestris
Endemic.

♂ adult

Greater Antillean Bullfinch
de: Rotsteiß-Gimpelfink
fr: Sporophile petit-coq
es: Come Ñame Violaceo
ja: クロアカウソ
cn: 大安德牛雀
Loxigilla violacea

ssp. ruficollis ♂ adult

Yellow-shouldered Grassquit
de: Goldbug-Gimpelfink
fr: Sporophile mantelé
es: Semillerito de Hombros Amarillos
ja: キゴロモコメワリ
cn: 黄肩草雀
Loxipasser anoxanthus
Endemic.

adult

New World Buntings and Sparrows - *Passerellidae*

The family of New World Buntings and Sparrows is restricted to America. They are relatively small and many are inconspicuously colored. They feed mainly on seeds.

Grasshopper Sparrow
 de: Heuschreckenammer
 fr: Bruant sauterelle
 es: Sabanero Chapulín
 ja: イナゴヒメドリ
 cn: 黄胸草鹀
Ammodramus savannarum

ssp. savannarum ♂ adult

Drawing P.H.Gosse

Spindalises - *Spindalidae*

The family of Spndalises has only been considered an independent family since 2013; they had previously been seen as representatives of the Tanagers. They occur in the Caribbean and on the island of Cozumel off the coast of Mexico. They are resident birds. They reach body lengths of 13 cm to 18 cm. They have short, powerful beaks and nine primary wing feathers. Their habitat is very diverse, they are primarily fruit-eaters and feed mainly on berries, but occasionally also on insects and other invertebrates. Their nests are cup-shaped.

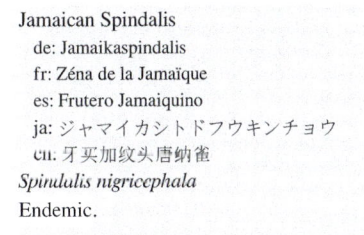

Jamaican Spindalis
 de: Jamaikaspindalis
 fr: Zéna de la Jamaïque
 es: Frutero Jamaiquino
 ja: ジャマイカシトドフウキンチョウ
 cn: 牙买加纹头唐纳雀
Spindulis nigricephala
Endemic.

♂ adult

Photo W.J.Daunicht

Cardinals - *Cardinalidae*

The family of Cardinals is common in North and South America. They are about 12 cm to 24 cm long. They have very powerful, conical beaks and nine hand wings; the plumage has mostly bright colors in red, yellow or blue. Their habitat are bush and forest landscapes. They feed mostly on seeds.

Summer Tanager
de: Sommertangare
fr: Tangara vermillon
es: Quitrique Colorado
ja: ナツフウキンチョウ
cn: 玫红丽唐纳雀
Piranga rubra

♂ breeding

Scarlet Tanager
de: Scharlachtangare
fr: Tangara écarlate
es: Quitrique Rojo
ja: アカフウキンチョウ
cn: 猩红丽唐纳雀
Piranga olivacea

♂ adult

Rose-breasted Grosbeak
de: Rosenbrust-Kernknacker
fr: Cardinal à poitrine rose
es: Piquigrueso Degollado
ja: ムネアカイカル
cn: 玫胸白斑翅雀
Pheucticus ludovicianus

♂ adult

Blue Grosbeak
 de: Azurbischof
 fr: Guiraca bleu
 es: Piquigrueso Azul
 ja: ルリイカル
 cn: 斑翅蓝彩鹀
Passerina caerulea

♂ adult

Indigo Bunting
 de: Indigofink
 fr: Passerin indigo
 es: Azulillo Norteño
 ja: ルリノジコ
 cn: 靛彩鹀
Passerina cyanea

♂ adult

Painted Bunting
 de: Papstfink
 fr: Passerin nonpareil
 es: Azulillo Pintado
 ja: ゴシキノジコ
 cn: 丽彩鹀
Passerina ciris
Near threatened.

♂ adult

Dickcissel
 de: Dickzissel
 fr: Dickcissel d'Amérique
 es: Arrocero Americano
 ja: ムナグロノジコ
 cn: 美洲雀
Spiza americana

adult

New World Blackbirds - *Icteridae*

The family of New World Blackbirds is restricted to North and South America. The species are very different in size: 19 cm to 55 cm. The plumage is often black with brightly colored areas. Their food is very varied. Some species are breeding parasites, while the other species nest very differently. The hanging nests are unusual.

Bobolink
de: Bobolink
fr: Goglu des prés
es: Soldadito Arrocero
ja: ボボリング
cn: 刺歌雀
Dolichonyx oryzivorus

♂ breeding

Orchard Oriole
de: Gartentrupial
fr: Oriole des vergers
es: Turpial de los Huertos
ja: アカクロムクドリモドキ
cn: 圃拟鹂
Icterus spurius

♂ adult

Jamaican Oriole
de: Jamaikatrupial
fr: Oriole de la Jamaïque
es: Turpial Caribeño
ja: ジャマイカムクドリモドキ
cn: 牙买加拟鹂
Icterus leucopteryx

ssp. leucopteryx adult

Baltimore Oriole
  de: Baltimoretrupial
  fr: Oriole de Baltimore
  es: Ictérido anaranjado
  ja: ボルチモアムクドリモドキ
  cn: 橙腹拟鹂
Icterus galbula

♂ adult

Jamaican Blackbird
  de: Bromelienstärling
  fr: Carouge de la Jamaïque
  es: Turpial Jamaiquino
  ja: ジャマイカクロムクドリモドキ
  cn: 牙买加黑鹂
Nesopsar nigerrimus
Endemic.
Endangered.

adult

Shiny Cowbird
  de: Seidenkuhstärling
  fr: Vacher luisant
  es: Vaquero Mirlo
  ja: テリバネコウウチョウ
  cn: 紫辉牛鹂
Molothrus bonariensis
Introduced.

♂ adult

Greater Antillean Grackle
  de: Antillengrackel
  fr: Quiscale noir
  es: Zanate Antillano
  ja: アンチルクロムクドリモドキ
  cn: 黑拟八哥
Quiscalus niger

ssp. crassirostris ♂ adult

Finches - *Fringillidae*

The family of Finches is widespread worldwide except for Australia and some oceanic islands. The body length is between 11 cm and 22 cm. They eat seeds and buds, insects almost only during the breeding season. The nest is built by the female from twigs, grass, moss and lichen in the form of a padded bowl.

Jamaican Euphonia
de: Gimpelorganist
fr: Organiste de la Jamaïque
es: Fruterito Jamaiquino
ja: ジャマイカスミレフウキンチョウ
cn: 牙买加歌雀
Euphonia jamaica
Endemic.

Sparrows - *Passeridae*

The sparrow family is native to Europe, Asia and Africa. However, one species managed to conquer the entire globe. The small birds are only 10 cm to 18 cm long. The conical beak indicates that they are grain eaters.

House Sparrow
de: Haussperling
fr: Moineau domestique
es: Gorrión Doméstico
ja: イエスズメ
cn: 家麻雀
Passer domesticus

www.avitopia.net/bird.en/?vid=6150202
www.avitopia.net/bird.en/?aud=6150202

Weavers - *Ploceidae*

The family of Weavers is widespread in Africa, a few species are found in Asia. The body sizes start with 10 cm and reach 70 cm in some species only for the males and only in breeding plumage. Weavers feed on seeds. They are usually gregarious birds and usually breed in colonies. The typical weaver's nest is a closed sphere that hangs from the end of a twig.

Yellow-crowned Bishop
- de: Tahaweber
- fr: Euplecte vorabé
- es: Obispo de Corona Amarilla
- ja: オウゴンチョウ
- cn: 黄顶巧织雀

Euplectes afer
Introduced.

www.avitopia.net/bird.en/?vid=6176506

♂ breeding

Waxbills - *Estrildidae*

The family of Waxbills is found in Africa, South Asia, and Australia. They are small birds with a body length of 9 cm to 14 cm. The beak is short and strong. Most of the species are grain eaters. As a rule, they are sociable birds. The nests are messy structures that are built by both parents. The chicks' throats are often very contrasting in color. Some species are ready for breeding after just a few months.

Scaly-breasted Munia
- de: Muskatamadine
- fr: Capucin damier
- es: Capuchino Nutmeg
- ja: シマキンパラ
- cn: 斑文鸟

Lonchura punctulata
Introduced.

adult

Black-headed Munia
de: Schwarzbauchnonne
fr: Capucin à dos marron
es: Monjita Tricolor
ja: ギンパラ
cn: 黑头文鸟
Lonchura malacca
Introduced.

Chestnut Munia
de: Schwarzkopfnonne
fr: Capucin à tête noire
es: Capuchino Castaño
ja: ミナミギンパラ
cn: 栗腹文鸟
Lonchura atricapilla
Introduced.

Index of English Names

Amazon
  Black-billed 63
  Yellow-billed 64
Anhinga 21
Ani
  Smooth-billed 52
Avocet
  American 32
Bananaquit 87
Becard
  Jamaican 68
Bishop
  Yellow-crowned 95
Bittern
  American 22
  Least 22
Blackbird
  Jamaican 93
Bobolink 92
Booby
  Brown 19
  Masked 19
  Red-footed 19
Bullfinch
  Greater Antillean 88
Bunting
  Indigo 91
  Painted 91
Canvasback 10
Caracara
  Northern Crested 62
Catbird
  Grey 76
Coot
  American 31
Cormorant
  Neotropic 20
Cowbird
  Shiny 93
Crake
  Uniform 29
  Yellow-breasted 29
Crow
  Jamaican 70
Cuckoo
  Black-billed 55

  Chestnut-bellied 53
  Jamaican Lizard 53
  Mangrove 52
  Yellow-billed 52
Dickcissel 91
Dove
  Caribbean 51
  Common Ground 50
  Eurasian Collared 49
  Mourning 51
  White-winged 51
  Zenaida 51
Dowitcher
  Short-billed 40
Duck
  Fulvous Whistling 8
  Masked 12
  Ring-necked 11
  Ruddy 12
  West Indian Whistling 8
  Wood 8
Egret
  Cattle 24
  Great 23
  Reddish 24
  Snowy 23
Elaenia
  Greater Antillean 65
  Jamaican 65
Euphonia
  Jamaican 94
Falcon
  Peregrine 63
Finch
  Saffron 87
Flamingo
  American 14
Flycatcher
  Acadian 66
  Great Crested 66
  Rufous-tailed 67
  Sad 66
  Stolid 67
Frigatebird
  Magnificent 18
Gadwall 9

gallinule
  American 30
Gallinule
  Common 31
Gannet
  Northern 20
Gnatcatcher
  Blue-grey 73
Godwit
  Hudsonian 36
  Marbled 37
Golden-Plover
  American 34
Grackle
  Greater Antillean 93
Grassquit
  Black-faced 88
  Yellow-faced 87
  Yellow-shouldered 88
Grebe
  Least 13
  Pied-billed 13
Grosbeak
  Blue 91
  Rose-breasted 90
Guineafowl
  Helmeted 12
Gull
  Bonaparte's 43
  Great Black-backed 44
  Herring 44
  Laughing 43
  Ring-billed 44
Hawk
  Broad-winged 28
  Red-tailed 29
  Sharp-shinned 28
Hen-harrier
  American 28
Heron
  Black-crowned Night 24
  Great Blue 22
  Green 24
  Little Blue 23
  Tricolored 23
  Yellow-crowned Night 25

Hummingbird
  Ruby-throated 59
  Vervain 60
Ibis
  American White 25
  Glossy 25
Jacana
  Northern 35
Jaeger
  Long-tailed 43
  Parasitic 42
  Pomarine 42
Kestrel
  American 62
Killdeer 35
Kingbird
  Eastern 67
  Grey 67
  Loggerhead 68
Kingfisher
  Belted 61
Kite
  Mississippi 28
  Swallow-tailed 27
Knot
  Red 37
Limpkin 31
Mango
  Jamaican 59
Martin
  Caribbean 71
  Purple 71
  Sand 72
Merganser
  Hooded 11
Merlin 63
Mockingbird
  Bahama 76
  Northern 76
Munia
  Black-headed 96
  Chestnut 96
  Scaly-breasted 95
Myna
  Common 77
Nighthawk
  Antillean 56
  Common 56

Noddy
  Black 45
  Brown 44
Orangequit 88
Oriole
  Baltimore 93
  Jamaican 92
  Orchard 92
Osprey 27
Ovenbird 79
Owl
  Barn 54
  Burrowing 54
  Jamaican 55
  Short-eared 55
  Stygian 55
Oystercatcher
  American 33
Parakeet
  Jamaican 64
Parrotlet
  Green-rumped 64
Parula
  Northern 83
Pauraque
  Jamaican 56
Pelican
  Brown 21
Petrel
  Black-capped 14
  Leach's Storm 16
  Wilson's Storm 16
Pewee
  Eastern Wood 65
  Jamaican 66
Phalarope
  Red-necked 40
Pigeon
  Common 48
  Plain 49
  Ring-tailed 49
  Scaly-naped 48
  White-crowned 49
Pintail
  Northern 10
  White-cheeked 10
Pipit
  Buff-bellied 78

Plover
  Grey 33
  Piping 35
  Semipalmated 34
plover
  Snowy 34
Plover
  Wilson's 34
Potoo
  Northern 57
Quail-Dove
  Blue-headed 50
  Crested 50
  Ruddy 50
Rail
  Black 29
  Clapper 30
Redhead 11
Redstart
  American 83
Ruff 37
Sanderling 38
Sandpiper
  Buff-breasted 39
  Least 38
  Pectoral 39
  Semipalmated 39
  Solitary 41
  Spotted 40
  Stilt 38
  Upland 36
  Western 39
  White-rumped 38
Sapsucker
  Yellow-bellied 62
Scaup
  Lesser 11
Shearwater
  Audubon's 15
  Cory's 15
  Manx 15
  Sooty 15
Shoveler
  Northern 9
Skimmer
  Black 48
Skua
  South Polar 42

Snipe
  Wilson's 40
Solitaire
  Rufous-throated 74
Sora 30
Sparrow
  Grasshopper 89
  House 94
Spindalis
  Jamaican 89
Spoonbill
  Roseate 26
Starling
  Common 77
Stilt
  Black-necked 32
Stork
  Wood 18
Streamertail 60
Swallow
  American Cliff 73
  Barn 72
  Cave 73
  Golden 72
  Northern Rough-winged 71
  Tree 72
Swift
  American Black 58
  Antillean Palm 59
  Chimney 58
  White-collared 58
Tanager
  Scarlet 90
  Summer 90
Teal
  Blue-winged 9
  Green-winged 10
Tern
  Arctic 47
  Black 46
  Bridled 45
  Common 46
  Forster's 47
  Gull-billed 46

Least 45
Roseate 46
Royal 47
Sandwich 47
Sooty 45
Thrush
  Grey-cheeked 74
  Swainson's 75
  White-chinned 75
  White-eyed 75
  Wood 75
Tody
  Jamaican 60
Tropicbird
  Red-billed 17
  White-tailed 17
Turnstone
  Ruddy 37
Veery 74
Vireo
  Black-whiskered 70
  Blue Mountain 69
  Jamaican 69
  Red-eyed 70
  Yellow-throated 69
Vulture
  Turkey 26
Warbler
  American Yellow 84
  Arrowhead 82
  Bay-breasted 84
  Black-and-white 80
  Black-throated Blue 85
  Black-throated Green 86
  Blackpoll 85
  Blue-winged 80
  Cape May 83
  Chestnut-sided 84
  Connecticut 81
  Golden-winged 80
  Hooded 83
  Kentucky 82
  Magnolia 84
  Mourning 82

Palm 85
Prairie 86
Prothonotary 81
Swainson's 81
Tennessee 81
Worm-eating 79
Yellow-rumped 85
Yellow-throated 86
Waterthrush
  Louisiana 79
  Northern 80
Waxwing
  Cedar 78
Whimbrel 36
Whip-poor-will
  Eastern 57
Wigeon
  American 9
Willet 41
Woodpecker
  Jamaican 61
Yellowlegs
  Greater 41
  Lesser 41
Yellowthroat
  Common 82

Index of Scientific Names

Accipiter 28
Acridotheres 77
Actitis 40
acuta, Anas 10
aethereus, Phaethon 17
afer, Euplectes 95
affinis, Aythya 11
agilis, Amazona 63
agilis, Oporornis 81
Aix 8
ajaja, Platalea 26
alba, Ardea 23
alba, Calidris 38
alba, Tyto 54
albus, Eudocimus 25
alcyon, Megaceryle 61
altiloquus, Vireo 70
Amaurolimnas 30
Amazona 63
americana, Aythya 11
americana, Fulica 31
americana, Mareca 9
americana, Mycteria 18
americana, Recurvirostra 32
americana, Setophaga 83
americana, Siphonorhis 56
americana, Spiza 91
americanus, Coccyzus 52
Ammodramus 89
anaethetus, Onychoprion 45
Anas 10
Anhinga 21
anhinga, Anhinga 21
ani, Crotophaga 52
Anous 44
anoxanthus, Loxipasser 88
Anthracothorax 59
Anthus 78
antillarum, Sternula 45
Antrostomus 57
Aramus 31
arborea, Dendrocygna 8
Archilochus 59
Ardea 22
Ardenna 15
Arenaria 37

argentatus, Larus 44
asiatica, Zenaida 51
Asio 55
Athene 54
atricapilla, Lonchura 96
atricilla, Leucophaeus 43
aura, Cathartes 26
aurantius, Turdus 75
aurita, Zenaida 51
aurocapilla, Seiurus 79
Aythya 10
bahamensis, Anas 10
barbirostris, Myiarchus 66
Bartramia 36
bassanus, Morus 20
bicolor, Dendrocygna 8
bicolor, Tachycineta 72
bicolor, Tiaris 88
Bombycilla 78
bonariensis, Molothrus 93
Botaurus 22
brasilianus, Phalacrocorax 20
Bubulcus 24
Buteo 28
Butorides 24
caerulea, Egretta 23
caerulea, Passerina 91
caerulea, Polioptila 73
caerulescens, Setophaga 85
Calidris 37
Calonectris 15
campestris, Euneornis 88
canutus, Calidris 37
Caracara 62
caribaea, Patagioenas 49
carolina, Porzana 30
carolinensis, Dumetella 76
castanea, Setophaga 84
Cathartes 26
Catharus 74
caudifasciatus, Tyrannus 68
cedrorum, Bombycilla 78
Chaetura 58
Charadrius 34
cheriway, Caracara 62

Chlidonias 46
Chordeiles 56
Chroicocephalus 43
chrysoptera, Vermivora 80
Circus 28
ciris, Passerina 91
citrea, Protonotaria 81
citrina, Setophaga 83
clypeata, Spatula 9
Coccyzus 52
Coereba 87
collaria, Amazona 64
collaris, Aythya 11
colubris, Archilochus 59
Columba 48
columbarius, Falco 63
Columbina 50
concolor, Amaurolimnas 30
Contopus 65
coronata, Setophaga 85
Corvus 70
cotta, Myiopagis 65
crecca, Anas 10
crepitans, Rallus 30
crinitus, Myiarchus 66
Crotophaga 52
cucullatus, Lophodytes 11
cunicularia, Athene 54
cyanea, Passerina 91
cyanocephala, Starnoenas 50
cyanoptera, Vermivora 80
Cypseloides 58
dactylatra, Sula 19
decaocto, Streptopelia 49
delawarensis, Larus 44
delicata, Gallinago 40
Dendrocygna 8
diomedea, Calonectris 15
discolor, Setophaga 86
discors, Spatula 9
Dolichonyx 92
domesticus, Passer 94
dominica, Pluvialis 34
dominica, Setophaga 86
dominicensis, Progne 71

dominicensis, Tyrannus 67
dominicus, Nomonyx 12
dominicus, Tachybaptus 13
dougallii, Sterna 46
Dumetella 76
Egretta 23
Elaenia 65
Elanoides 27
Empidonax 66
erythropthalmus, Coccyzus 53
euchrysea, Tachycineta 72
Eudocimus 25
Euneornis 88
Euphonia 94
Euplectes 95
Eupsittula 64
exilis, Ixobrychus 22
falcinellus, Plegadis 25
Falco 62
fallax, Elaenia 65
fedoa, Limosa 37
flammeus, Asio 55
flaveola, Coereba 87
flaveola, Sicalis 87
flavifrons, Vireo 69
flavipes, Tringa 41
flaviventer, Hapalocrex 29
forficatus, Elanoides 27
formosa, Geothlypis 82
Forpus 64
forsteri, Sterna 47
Fregata 18
Fulica 31
fulva, Petrochelidon 73
fuscatus, Onychoprion 45
fuscescens, Catharus 74
fuscicollis, Calidris 38
galbula, Icterus 93
galeata, Gallinula 31
Gallinago 40
Gallinula 31
Gelochelidon 46
genibarbis, Myadestes 74
Geothlypis 82
Geotrygon 50
grammicus, Pseudoscops 55
grisea, Ardenna 15
griseus, Limnodromus 40

guarauna, Aramus 31
gundlachii, Chordeiles 56
gundlachii, Mimus 76
haemastica, Limosa 36
Haematopus 33
haliaetus, Pandion 27
Hapalocrex 29
hasitata, Pterodroma 14
Helmitheros 79
herodias, Ardea 22
Himantopus 32
himantopus, Calidris 38
Hirundo 72
hirundo, Sterna 46
hudsonius, Circus 20
Hylocichla 75
ibis, Bubulcus 24
Icterus 92
Ictinia 28
inornata, Patagioenas 49
interpres, Arenaria 37
Ixobrychus 22
Jacana 35
jamaica, Euphonia 94
jamaicensis, Buteo 29
jamaicensis, Corvus 70
jamaicensis, Laterallus 29
jamaicensis, Leptotila 51
jamaicensis, Nyctibius 57
jamaicensis, Oxyura 12
jamaicensis, Turdus 75
Larus 44
Laterallus 29
lentiginosus, Botaurus 22
Leptotila 51
lepturus, Phaethon 17
leucocephala, Patagioenas 40
leucogaster, Sula 19
Leucophaeus 43
leucopteryx, Icterus 92
leucorhoa, Oceanodroma 16
lherminieri, Puffinus 15
Limnodromus 40
Limnothlypis 81
Limosa 36
livia, Columba 48
lobatus, Phalaropus 40
Lonchura 96

longicauda, Bartramia 36
longicaudus, Stercorarius 43
Lophodytes 11
Loxigilla 88
Loxipasser 88
ludovicianus, Pheucticus 90
maccormicki, Stercorarius 42
macroura, Zenaida 51
macularius, Actitis 40
magnificens, Fregata 18
magnolia, Setophaga 84
malacca, Lonchura 96
mango, Anthracothorax 59
Mareca 9
marinus, Larus 44
martinica, Porphyrio 30
mauri, Calidris 39
maximus, Thalasseus 47
Megaceryle 61
Melanerpes 61
melanoleuca, Tringa 41
melanotos, Calidris 39
meleagris, Numida 12
Mellisuga 60
melodus, Charadrius 35
mexicanus, Himantopus 32
Mimus 76
minima, Mellisuga 60
minimus, Catharus 74
minor, Chordeiles 56
minor, Coccyzus 52
minutilla, Calidris 38
minutus, Anous 45
mississippiensis, Ictinia 28
Mniotilta 80
modestus, Vireo 69
Molothrus 93
montana, Geotrygon 50
Morus 20
motacilla, Parkesia 79
mustelina, Hylocichla 75
Myadestes 74
Mycteria 18
Myiarchus 66
Myiopagis 65
nana, Eupsittula 64
Nesopsar 93
niger, Chlidonias 46
niger, Cypseloides 58

niger, Pachyramphus 68
niger, Quiscalus 93
niger, Rynchops 48
nigerrimus, Nesopsar 93
nigricephala, Spindalis 89
nilotica, Gelochelidon 46
nivosus, Charadrius 34
Nomonyx 12
noveboracensis, Parkesia 80
Numenius 36
Numida 12
Nyctanassa 25
Nyctibius 57
Nycticorax 24
nycticorax, Nycticorax 24
occidentalis, Pelecanus 21
oceanicus, Oceanites 16
Oceanites 16
Oceanodroma 16
olivacea, Piranga 90
olivaceus, Tiaris 87
olivaceus, Vireo 70
Onychoprion 45
Oporornis 81
Oreothlypis 81
oryzivorus, Dolichonyx 92
osburni, Vireo 69
Oxyura 12
Pachyramphus 68
palliatus, Haematopus 33
pallidus, Contopus 66
palmarum, Setophaga 85
Pandion 27
paradisaea, Sterna 47
parasiticus, Stercorarius 42
Parkesia 79
Passer 94
Passerina 91
passerina, Columbina 50
passerinus, Forpus 64
Patagioenas 48
pelagica, Chaetura 58
Pelecanus 21
pensylvanica, Setophaga 84
peregrina, Oreothlypis 81
peregrinus, Falco 63
petechia, Setophaga 84
Petrochelidon 73
phaeopus, Numenius 36

Phaethon 17
Phalacrocorax 20
Phalaropus 40
pharetra, Setophaga 82
Pheucticus 90
philadelphia, Chroicocephalus 43
philadelphia, Geothlypis 82
phoenicobia, Tachornis 59
Phoenicopterus 14
Piranga 90
Platalea 26
platypterus, Buteo 28
Plegadis 25
Pluvialis 33
pluvialis, Coccyzus 53
podiceps, Podilymbus 13
Podilymbus 13
Polioptila 73
polyglottos, Mimus 76
polytmus, Trochilus 60
pomarinus, Stercorarius 42
Porphyrio 30
Porzana 30
Progne 71
Protonotaria 81
Pseudoscops 55
Pterodroma 14
Puffinus 15
puffinus, Puffinus 15
pugnax, Calidris 37
punctulata, Lonchura 95
pusilla, Calidris 39
pyrrhonota, Petrochelidon 73
Quiscalus 93
radiolatus, Melanerpes 61
Rallus 30
Recurvirostra 32
Riparia 72
riparia, Riparia 72
ruber, Phoenicopterus 14
rubescens, Anthus 78
rubra, Piranga 90
rufescens, Egretta 24
rustica, Hirundo 72
ruticilla, Setophaga 83
Rynchops 48
sandvicensis, Thalasseus 47
savannarum, Ammodramus 89

Seiurus 79
semipalmata, Tringa 41
semipalmatus, Charadrius 34
serripennis, Stelgidopteryx 71
Setophaga 82
Sicalis 87
Siphonorhis 56
solitaria, Tringa 41
sparverius, Falco 62
Spatula 9
Sphyrapicus 62
Spindalis 89
spinosa, Jacana 35
Spiza 91
sponsa, Aix 8
spurius, Icterus 92
squamosa, Patagioenas 48
squatarola, Pluvialis 33
Starnoenas 50
Stelgidopteryx 71
Stercorarius 42
Sterna 46
Sternula 45
stolidus, Anous 44
stolidus, Myiarchus 67
strepera, Mareca 9
Streptopelia 49
Streptoprocne 58
striata, Setophaga 85
striatus, Accipiter 28
Sturnus 77
stygius, Asio 55
subis, Progne 71
subruficollis, Calidris 39
Sula 19
sula, Sula 19
swainsonii, Limnothlypis 81
Tachornis 59
Tachybaptus 13
Tachycineta 72
Thalasseus 47
thula, Egretta 23
Tiaris 87
tigrina, Setophaga 83
Todus 60
todus, Todus 60
trichas, Geothlypis 82
tricolor, Egretta 23

Tringa 41
tristis, Acridotheres 77
Trochilus 60
Turdus 75
Tyrannus 67
tyrannus, Tyrannus 67
Tyto 54
ustulatus, Catharus 75
validus, Myiarchus 67
valisineria, Aythya 10
varia, Mniotilta 80

varius, Sphyrapicus 62
Vermivora 80
vermivorum, Helmitheros 79
versicolor, Geotrygon 50
vetula, Coccyzus 53
violacea, Loxigilla 88
violacea, Nyctanassa 25
virens, Contopus 65
virens, Setophaga 86
Vireo 69
virescens, Butorides 24

virescens, Empidonax 66
vociferus, Antrostomus 57
vociferus, Charadrius 35
vulgaris, Sturnus 77
wilsonia, Charadrius 34
Zenaida 51
zonaris, Streptoprocne 58

Additional Copyright Terms

Please note the following copyright terms applying to the individual fotos, drawings, and grafics contained in this e-book:

As far as known, the names of the authors (photographers, graphic artists and other creators) of the individual photos, drawings, graphics, maps and other works are placed directly next to the respective images (Attribution).

Moreover, every photo, drawing, graphic, map and each other work is labeled with an abbreviation that refers to the license under which the work has been reproduced here. These abbreviations mean:

| AU | AVITOPIA holds the copyright. All rights reserved. |

| LIC | AVITOPIA has obtained certain licenses from the holder of the rights of use. All rights reserved. |

| S2.0 | To the best of the knowledge of AVITOPIA this image is subject to the Creative Commons Attribution Share Alike License 2.0 the complete conditions of which can be found at Creativecommons.org/licenses/by-sa/2.0/deed.en . The author has some rights reserved. |

| S2.5 | To the best of the knowledge of AVITOPIA this image is subject to the Creative Commons Attribution Share Alike License 2.5 the complete conditions of which can be found at Creativecommons.org/licenses/by-sa/2.5/deed.en . The author has some rights reserved. |

| S3.0 | To the best of the knowledge of AVITOPIA this image is subject to the Creative Commons Attribution Share Alike License 3.0 the complete conditions of which can be found at Creativecommons.org/licenses/by-sa/3.0/deed.en . The author has some rights reserved. |

| S4.0 | To the best of the knowledge of AVITOPIA this image is subject to the Creative Commons Attribution Share Alike License 4.0 the complete conditions of which can be found at Creativecommons.org/licenses/by-sa/4.0/deed.en . The author has some rights reserved. |

| A2.0 | To the best of the knowledge of AVITOPIA this image is subject to the Creative Commons Attribution License 2.0 the complete conditions of which can be found at Creativecommons.org/licenses/by/2.0/deed.en . The author has some rights reserved. |

| A3.0 | To the best of the knowledge of AVITOPIA this image is subject to the Creative Commons Attribution License 3.0 the complete conditions of which can be found at Creativecommons.org/licenses/by/3.0/deed.en . The author has some rights reserved. |

| PD | To the best of the knowledge of AVITOPIA this image is in the public domain. However, AVITOPIA does not accept any liability if you use such an image. It is the obligation of the user, to verify the absence of any right reservations under his/her jurisdiction. |

Printed in Great Britain
by Amazon